BASICS OF ENTOMOLOGY

BASICS OF ENTOMOLOGY

Gyan Deep Singh

ANMOL PUBLICATIONS PVT. LTD.
NEW DELHI - 110 002 (INDIA)

ANMOL PUBLICATIONS PVT. LTD.
H.O.: 4374/4B, Ansari Road, Daryaganj,
New Delhi-110 002 (India)
Ph.: 23278000, 23261597
B.O.: No. 1015, Ist Main Road, BSK IIIrd Stage
IIIrd Phase, IIIrd Block,
Bangalore - 560 085 (India)
Visit us at: www.anmolpublications.com

Basics of Entomology

First Edition, 2008
ISBN 978-81-261-3729-9

PRINTED IN INDIA

Printed at Mehra Offset Press, Delhi.

Contents

Preface

Entomology is attracting so much interest among all classes, and has become such a favourite pursuit, it appeared to the author of the present work that there was room for a book of moderate size, in which the most typical and remarkable insects of all orders, and from all parts of the world, should be popularly described and figured. Previous works on entomology have generally treated only of a single order of insects, or else of British insects.

This book is chiefly intended as a compressed popular Introduction to the classification of insects in general, and by the careful omission of all technicalities not absolutely necessary for this object, sufficient space has been gained to give a short and readable, though necessarily somewhat condensed, account of all the families of insects, interspersed with occasional observations on their habits and economic importance, preference being given to those which have not previously appeared in English popular works.

Author

Chapter 1

Introduction

Entomology is the study of insects, the most abundant form of animal life on earth. Three quarters of a million species of insect have already been described and estimates for the number of species awaiting discovery range from 1 million to 30 million. So for any budding entomologists out there, you have an excellent chance of discovering a new species.

Given that there are so many species of insect, it follows that insects are a major component of the earth's biodiversity. They inhabit every terrestrial and freshwater ecosystem and by studying the ecological roles insects play we can have a better understanding of how those ecosystems function. Insects have important roles as plant consumers and herbivores, a food source for other organisms, scavengers and detritivores, predators and parasites. Insects also directly affect human welfare by competing with us for food and transmitting diseases. However, not all insects are detrimental to human welfare. For example, bees are used to produce honey, silkworms to produce silk and many predatory species are used to control crop pests.

Insects have incredibly diverse morphological, physiological, and behavioural adaptations to their surroundings which makes the study of insects a fascinating subject. Many features of insect biology also make them ideal to use as model biological systems. Their abundance, short life cycle, reproductive potential and small size allow scientific experiments to be set up, monitored and duplicated with relative ease in almost any location. Much of our basic understanding of genetics, population ecology, and evolution has resulted from experimentation with insects.

ENERGY METABOLISM

Muscle contraction, ion transport, anabolic reaction pathways, and most other forms of cellular work are all powered by hydrolysis of adenosine triphosphate (ATP). The phosphodiester linkages in ATP are high energy bonds, and the potential chemical energy stored in ATP is released during the enzymatic hydrolysis of these phosphate bonds. The released energy is captured and utilized to power cell work. ATP is nearly ubiquitous, and is generally regarded as a sort of universal currency in cellular energetics. ATP is an unstable molecule, and large amounts can not be accumulated in anticipation of cellular activity. Hence, ATP is often generated during activity, especially energy-intensive activities such as insect flight.

ATP is part of the adenylate system. Adenosine monophosphote (AMP) and adenosine diphosphate (ADP) complete the system. These three molecules, AMP, ADP and ATP, store and transduce most of the energy used in cellular metabolism. The relative amounts of each make up what we call the "energy charge" of any cell. If all adenosine in the adenylate system were present as ATP, the energy charge would be 1.0; if all adenosine were present as AMP the energy charge would be 0.0. Cells generally maintain a homeostasis of energy charge.

When the energy charge declines below a set point appropriate to any given cell, there follows a shift toward ATP-generating pathways. On the other hand, when the energy charge is high, cells favour ATP-consuming pathways. Homeostasis of energy charge results from balancing ATP-generating and ATP-consuming pathways. Just to give you a number, the energy charge of most cells is about 0.85, perhaps something higher than we might expect.

Let us wonder in a little bit more detail how cells balance ATP-generating and ATP-consuming pathways. We gain a bit more understanding by regarding the biological significance of the adenylate system. The adenylate system influences cellular metabolism in two ways. First, as just mentioned, the relative amounts of each adenosine species determines the energy charge of cells. Second, key enzymes in some metabolic pathways are regulated by the instantaneous concentrations of ATP and AMP. Hence, the adenylate system is both a regulatory and and energy-yielding entity.

Respiration, or gas exchange, provides oxygen to mitochondria to support the rapid generation of ATP during energy intensive activities, such as flight. Mitochondria, and to some extent, extra-mitochondrial cellular enzymes generate ATP by the step-wise oxidation of chemical fuels. Because the oxidation takes chemical fuels through several intermediary steps, this process is often called intermediary metabolism.

What are the flight fuels in insects? There are three major flight fuels. Honeybees and bumble bees, for examples, oxidize carbohydrate almost exclusively during flight. The most common carbohydrate for bees is the simple sugar glucose.

Some butterflies and moths rely for the most part on lipids, commonly known as fats, to support flight. Other insects, such as the tsetse and the Colorado potato beetle mostly re;u on one of the 20 amino acids - proline - during flight. Most insects are probably capable of using more than one energy substrate during various of their activities. For example, the desert locust usually begins a long flight powered by glucose metabolism; after about 15 to 20 minutes there is a shift to lipid metabolism for the duration of a long migratory flight.

Knowing the basic fuels used by most insects, we can next ask what are the usual storage forms of metabolic fuels? Carbohydrates are stored as glycogen. Glycogen is nothing more than a string of monomeric glucose units hooked up in a chain. Branches occur about every 8 to 12 glucose units in a glycogen macromolecule. The branching provides additional sites for rapid metabolism of glycogen: it can be elongated by adding simples sugars, and rapidly shortened by hydrolyzing them. The branching also increases solubility of the macromolecule. In vertebrate liver glycogen collects into dark-staining granules. Such granules have also been seen in preparations of insect fat bodies. Here we note that the fat body is not, as the name suggests, simply a fat storage depot. Rather, it is a complex metabolic centre more comparable to the vertebrate liver. Most glycogen is stored in the fat body and to a lesser extent is found in flight muscle cells.

The next major fuel - lipid - is stored as triacylglycerol. This is made of a glycerol backbone, which is a 3-carbon unit. The backbone may have 1, 2 or 3 fatty acids bonded in ester linkage. The nomenclature, then, is monoacyl-, diacyl-, and triacylglycerol, depending upon the number of fatty acids bonded in ester linkage to the backbone. Most lipid is stored in the fat body, although small amounts are found in other cells.

Some insects oxidize the amino acid proline, sometimes expressed in the correct nomenclature as an imino acid. Insect hemolymph is quite high in amino acid content - much higher than in mammalian blood. The hemolymph is the major storage site of proline. Proline is unique among the energy substrates because its storage, transport and metabolic forms are all the same.

Given that we have some idea of the major metabolic fuels and their storage sites and forms, we can go on with the observation that, by and large, the fat body is the major site of energy storage. The fat body is relatively distant from the site of utilization inside flight muscles and other insect tissues. This implies that the fuels must be transported. During flight, for example, fuels are transported mainly from the fat body, found mostly in the abdomen, to the thoracic muscles. This is one of the major physiological roles of hemolymph: the transport of nut-rients. The next general question, then, is what are the major transport forms of chemical energy?

Again, we start with carbohydrates, generally stored as glycogen in the fat body. Glycogen can be shortened, one glucose unit at a time. The released glucose units can then be coupled into a molecule of TREHALOSE. Trehalose is the principle blood sugar in insects, and it plays a comparable role to blood glucose in mammals. Trehalose is a disaccharide of two glucose units. Once formed, it is released from the fat body into the hemolymph.

In the hemolymph of blowflies, for an example, trehalose can be as concentrated to as much as 20 mg/ml. For comparison, human blood sugar is usually about 100 mg/ decaliter or 1 mg/ml. Due to the high hemolymph concentration, trehalose diffuses from the hemolymph into the muscle cells and other tissues where it is hydrolysed into 2 glucose units. The high concentration of trehalose in the hemolymph suggests that upon closer inspection there are two storage forms and two storage sites for carbohydrates. Glycogen is stored in the fat body and trehalose represents another, hemolymph, storage form.

The physiology of lipids, which we saw to be mainly stored in the fat body as triacylglycerol, is somewhat different. Unlike sugars, which are soluble in rather high concentrations in hemo-lymph, lipids are functionally defined as being insoluble in aqueous media. Because of this physicochemical property, the predominant lipid transport form is as diacylglycerol associated with a specific hemolymph lipoprotein called LIPOPHOREN. The most recent thinking, in the absence of real knowledge, is that fat-ty acids are hydrolysed from the glycerol backbone at the muscle cell membrane, then taken into the muscle while the lipophoren and glycerol backbone are recirculated back to the fat body.

We have come to the point of seeing what the major insect metabolic fuels are, their storage and transport forms and the main depots of storage. For the most part, energy metabolism in insects is not much different from other animals. A couple of major differences are transport of diacylglycerol and use of trehalose as a blood sugar.

	Carbohydrate	Lipid	Amino Acid
Storage Site	Fat body/blood	Fat body	hemolymph
Storage Form	Glycogen/trehalose	Fat body	Imino acid
Transport Form	Trehalose	diacylglycerol	Imino acid
Metabolic Form	Glucose	Fatty acid	Imino acid

The next questions are what are the major metabolic pathways for each of these forms of chemical energy? And, where do the enzymes in these pathways occur? These questions are best taken together.

Glucose is broken down to pyruvate, or pyruvic acid, by way of the Glycolytic or Embden-meyerhof pathway. Depending on what biology or biochemistry text you happen to look at, there is a lot of noise in the representation of this pathway, so we will focus on the key features. These are that glucose enters the pathway, and pyrvate is the end product. There is a net gain of two ATPs per glucose molecule. In short, glucose is a 6-carbon sugar that gets broken down into two 3-carbon compounds. The enzymes of this pathway occur in the cytoplasm of cells, outside the mitochondria. In mere point of interest, speaking in terms of evolution, this is probably the oldest energy-yielding pathway, and there are numerous life forms that rely solely on this system.

Much more energy, a total of 34 ATPs, depending on how you add them up, is obtained from the complete oxidation of glucose to carbon dioxide and water. This more complete breakdown is accomplished by enzymes of the Citric Acid or Kreb's Cycle. The key features are (one) that acetyl Coenzyme A, formed by the decarboxylation of pyruvic acid, enters the cycle by joining the 2-carbon acetyl unit to a 4-carbon compound to create the 6-carbon citric acid and (two) there is no real end point because of the cyclic nature of the pathway. Rather, carbon dioxide is created at two points in the cycle by decarboxylations - recall that in organismal respiration carbon dioxide is removed by gas exchange - and electron carriers become loaded with electrons at 4 points in the cycle. Finally, the equivalent of 1 ATP is generated with each round of the cycle. The enzymes for the decarboxylation of pyruvate to acetyl CoA and the enzymes of the citric acid cycle are all located in the matrix of the mitochondria.

The final step in the complete oxidation of glucose is that the loaded electron carriers dump their electrons onto the Electron Transport Chain. In this system, electrons flow from the carriers, NADH and FADH2, through a series of electron carriers, finally combining with oxygen to form water. At three points along this chain a single ATP can be created for each electron pair that flows down the chain. Now, turning back to the Kreb's cycle, we can see that each turn of the cycle yields 12 ATPs - three each for the NADH (9), two for the FADH (11) and one for the GTP (12). Thus, Kreb's cycle and the ETC operate in concert to complete the oxidation of sugar to carbon dioxide and water, with generation of considerable amounts of ATP. Whereas the enzymes of the citric acid cycle are located in the matrix of the mitochondria, the ETC is buried within the inner mitochondrial membrane.

Let us consider the oxidation of fatty acids. Once transported across the cell membrane into the cell, the fatty acids are hooked up to CoA, making fatty acyl Coenzyme A, which can be transferred to fatty acyl carnitine. This is simply a transfer step to get the fatty acid into a form that can be transported into the matrix of the mitochondria where it is again hooked up to CoA.

The carnitine, then diffuses back out of the mitochondrial space where it can reserve as a fatty acid carrier. Within the mitochondrial matrix, the fatty acyl CoA enters what is known as the ß-oxidation cycle. This pathway runs in circles, like most of us, and essentially repeatedly breaks 2-carbon units off the long-chain fatty acid which become acetyl Co-enzyme A. The acetyl CoA subsequently enters the Kreb's cycle and ETC where it undergoes complete oxidization to carbon dioxide and water. Recalling that one turn of Kreb's cycle yields 12 ATPs in concert with the ETC, we see that the 8 2-carbon units from a 16-carbon fatty acid can provide a considerable amount of energy for cell work.

The final substrate in our story will be the amino acid, proline. In the tsetse proline is oxidized in the sarcosomes, or mitochondria, according to this expression:

$$\text{Proline} + 21/2\ O_2 + 14\ \text{ADP} \rightarrow \text{Alanine} + H_2O + 2CO_2 + 14\ \text{ATP}$$

In essence, the nitrogen is taken off of the skeleton and the carbon portion is oxidized to carbon dioxide and water by entering the Kreb's cycle at a point beyond citric acid.

Let us have a look at the biological significance of sugars, lipids and proline as metabolic substrates. With regard to proine, we really have two questions. Why use an amino acid at all when it can be converted to carbohydrate which combines high solubility, good energy yield and convenient storage in hemolymph as trehalose and in fat body as glycogen? We understand proline metabolism best in the tsetse where it is part of the routine diet due to its blood-feeding habit. Bearing in mind that each step of any metabolic conversion results in some lose of energy, the idea is that it is more efficient to oxidize the dietary proline directly rather than convert it to something else. This is because the amino nitrogen atom must itself be metabolized through interconversion of other amino acids. It turns out that these other compounds are themselves better candidates for conversion to sugar. Then our second question would be why oxidize proline as opposed to some other amino acid? There are two main points.

One is the net yield of ATP per gram amino acid and the other is the solubility of the amino acids in aqueous hemolymph. On the basis of energy yield, probably only isoleucine and leucine are serious rivals to proline, with a possible argument for valine. But of these four amino acids, the solubility of all but proline is too low for maintenance of high concentrations with hemolymph. This argues that proline is probably the most suitable amino acid for energy substrate. And, again, may be quite appropriate when it is high in the diet.

With respect to flight duration, animals that can oxidize lipid as a flight fuel are capable of making long-range flights, whereas those that utilize carbohydrate are capable of only relatively short periods of continuous flight. This follows from the higher energy content of a given weight of lipid compared to sugar. Also, gly-cogen, the main carbohydrate store, is always stored with an appreciable amount of water associated with it. We can calculate that 1 mg of lipid can yield as much metabolic energy as about 8 mg of stored glycogen: thus, for long flights the weight economy of a fuel becomes important. Also, lipid produces more metabolic water than does carbohydrate and this may be important in the water relations of some insects. This reasoning is not limited to insects. Many migratory birds similarly accumulate lipid reserves to fuel long distance flights.

On the other hand, carbohydrate offers advantages to creatures like bumblebees. A key advantage is that the energy from glucose can be realised much faster than the energy from a fatty acid. It is generally true that the most rapidly metabolizing insects oxidize sugar.

FLIGHT MUSCLE PHYSIOLOGY

The ability to move is one of the basic characteristics of living things. To make this point, let us consider a number of different sorts of movement. Most of us are intimately familiar with more or less large scale movements, such as walking or running or watching birds and insects fly. We are also aware of a good number of more subtle movements, such as our heart-beats and movements of our viscera, actions that occur within organisms.

Movements also occur at the cellular level. Unicellular organisms are able to move themselves around, by amoeboid movement or the actions of flagella and cilia. Flagella and cilia are also found in certain cells, sperm for example, within metazoan animals. Movement takes place within cells, as well. Chromosomes are moved within cells by contractile elements of a spindle apparatus. Nutriments and other materials are transported within cells, down the length of a nerve axon, for example.

All movements are driven by variations on two general systems of proteins. One is the tubulin-dynein microtubular system, mostly seen within cells, and the other is the actin-myosin filament system of muscles. From the point of view of a comparative physiologist, muscle, and its actin-myosin contractile elements is a special tissue because no other tissue so clearly illustrates the idea of a common mechanism with so very many variations adapted to specific functions.

Turning our attention to insects, one of the most obvious features of insects is that most of them are able to fly. The common names of insects reflect flight again and again, for examples, dragonflies, damselflies, houseflies, mayflies, stoneflies, fruitflies, hoverflies, butterflies. I am sure you can think of many more examples. All bespeak the point that most insects are fliers. We can fairly say that most of the world's animals fly.

Flight bears on nearly every aspect of insect biology. Foraging for food, searching for mates, escape from predators or fly swatters, avoiding harsh or unfavourable conditions, and locating egg-laying sites are just a few points of insect biology intimately connected to flight behaviour. It follows that understanding various behavioural, aerodynamic, physiological, biochemical and ecological aspects of insect flight is valuable in understanding the nature of insects.

We first consider structures of insect flight muscles. Flight muscles are arranged into three broad categories based on structural grounds. In tubular muscles, fibrils are arranged radially about a column of sarcoplasm that contains several nuclei. Mitochondria are interspersed in the radial area of the muscle. Dragonflies feature this type of flight muscle structure. Tubular muscles occur in the legs of most insect species. Close-packed muscles are found in Lepidoptera and Orthoptera; people who get their security from scientific idiom call these microfibrillar muscles. In close-packed muscles large mitochondria are dispersed between the cylindrical fibrils, which are about 1.5 to 2.0 μm in diameter. Fibrillar muscles are characteristic of Hymenoptera, Diptera, Coleoptera and true bugs. These muscles have large fibriles, up to 5 μm in diameter, with huge mitochondria and nuclei scattered about.

Even though different phylogenic groups have developed various structural types of flight muscles, there are major similarities among them. All the muscle types have fibres as their physiological units because the fibres are bounded by a continuous sarcolemma or muscle cell membrane with the basement membrane attached. There are descriptions of fibres composed of several cells, however, a fibre is more correctly a single multinucleated cell, or syncitium. Also, the contractile proteins within the fibrils appear to be in similar arrangement.

Now, before we go forward, let us take a minute to clarify some language. People began looking closely at muscles during the 19th century, before articulation of a general cell theory. A consequence of this is a history of set of parallel nomenclatures, one for muscles and another for most other cell types. The terms for muscle usually include the prefix sarc- or my-, from Greek for flesh and muscle. So what we usually call the cell membrane or plasmalemma is called the sarcolemma in muscle. Similarly, mitochondria are called sarcosomes, the endoplasmic reticulum is called sarcoplasmic reticulum and cytosol is referred to as sarcoplasm. Still on nomenclature, muscles are comprised of fibres, which are made up of fibrils within which are the contractile filaments.

The ultrastructural arrangement of muscle is as follows. filaments are made up of the contractile proteins within muscle fibrils. Muscle fibres contain groups of fibrils; muscles are made of groups of fibres.

The most widely accepted mechanism of movement is known as the sliding filament theory of muscle contraction, first proposed in 1954. Each contractile unit is called a sarcomere, bounded on either end by Z-lines or discs. The area between Z-discs is sometimes called a Z-band. Other landmarks are the A-bands, the H-zone and the I-band. These bands are visible in microscopic preparations and are associated with the presence or absence of the major contractile proteins, actin and myosin. These terms date back to very early microscopic observation of muscles. The A of A-band stands for ANISOTROPIC, and the I of I-band stands for ISOTROPIC. The isotropic bands have uniform indices of refraction in microscope preparations. Consider a cross-sectional representation of muscle fibrils where thin actin filaments are arrayed around thicker myosin filaments. Try to get into the habit of saying thick myosin and thin actin. It may be easier to keep the names straight that way. In flight muscles, the actin filaments are very often found in a hexogonal array around the myosin filaments while in slow twitch visceral muscles the actin is in a circular arrangement. A last point here is that all insect flight muscles are striated. Variations in the register of the Z-discs within a muscle make this feature more or less apparent in muscles from insects of different orders.

In addition to actin and myosin filaments, other proteins are present in relatively small quantities. These include troponin and tropomyosin, collectively called native tropomyosin. The native tropomyosin is involved in regulation of the muscle contraction.

We know, now, that the physiological unit of contraction, the muscle fibre, is invested by the sarcolemma, or functional cell membrane. At frequent intervals this membrane is invaginated to form a system of transverse T-tubules. The invaginated transverse T-tubules lie in close association with vesicles of sarcoplasmic reticulum. Additional invaginations of the sarcolemma allow tracheoles to be very close to the sarcosomes which consume amazing amounts of oxygen during in-flight energy production. Note also that still more invaginations allow motor axons to have multiple synapses with the muscle fibre. This is an important point because unlike vertebrate muscles, the muscle membranes of insects does not propagate action potentials very. There is a requirement, then, for synaptic junctions about every 100 μm or so.

Summarizing muscle structure, the three structural types of muscle, tubular, close-packed and fibrillar, have in common the basic cellular organization of contractile proteins that present a more or less visible striation. Various invaginations for neuromuscular junctions, transverse T-tubule systems and tracheole position appear in all muscle membranes. We can now move on to physiology of muscle contraction.

A motor axon impulse is transmitted across a neuromuscular synapse to the muscle cell membrane resulting in a depolarization which spreads to the interior of the muscle fibre by the transverse T-tubule system. This is very well developed in insect flight muscles. Depolarization of the transverse T-tubule system causes release of calcium from vesicles of the sarcoplasmic reticulum. The calcium activates myosin ATPase, resulting in the two type of filaments sliding past each other. What we want to see is that neither the thin actin filaments nor the thick myosin filaments shorten during contraction. Rather, the Z-bands become increasingly shorter as the two types of filaments overlap each other more and more.

There is a great deal of experimental evidence on muscle contraction. If calcium is removed from an *in vitro* preparation of locust leg muscle, for example by addition of EGTA which selectively removes divalent cations, the muscle can not be stimulated to contract. Adding calcium chloride back into the buffer restores the ability to contract. Furthermore, calcium is resequestered by vesicles of the sarcoplasmic reticulum in the presence of ATP. The requirement for ATP shows that uptake of calcium requires energy: that is, it takes energy to relax.

The role of calcium in the interaction between thin actin and thick myosin filaments that result in muscle contraction is modulated through native tropomyosin. Troponin inhibits the actin-myosin interactions. This inhibition is calcium dependent due to the presence of tropomyosin. Thus, the contractile process involves an impulse from a motor axon which depolarizes the post-synaptic muscle cell membrane; the depolarization sweeps inward by way of the transverse T-tubules, resulting in release of calcium from vesicles of sarcoplasmic reticulum. Interaction of calcium and tropomyosin, which has specific calcium receptors, temporarily breaks the inhibition of actin-myosin interaction by troponin so that the two main contractile proteins slide past each other with the concomitant expenditure of ATP. The contraction cycle is concluded with the repolarization of the muscle cell membrane, the resequesterization of calcium by sarcoplasmic reticulum and relaxation of the muscle. Please note that relaxation is not the same thing as stretching the muscle to its uncontracted length. Most skeletal muscle systems in vertebrates and invertebrates occur in antagonistic pairs: following relaxation, a flight muscle is restretched by its pair-mate.

Let's go back to the idea that the two main contractile proteins slide past each other with the expenditure of ATP. We want to consider how the hydrolysis of ATP is translated into directional movement. We begin by appreciating a few points about myosin. Thick myosin filaments have a certain springiness or elasticity. Within a single filament, this allows some myosin heads can be attached to thin actin filaments while others are not attached. Each myosin filament has something in the range of 500 myosin heads, and each myosin head is able to bind and hydrolyze ATP. The process of actin and myosin filaments sliding past each other results from the combined actions of all the myosin heads.

Myosin heads move actin filaments in a cyclic process. A free myosin head can bind a molecule of ATP, the immediately hydrolyze the ATP into ADP and inorganic phosphate. Both products remain bound to the myosin head, and the energy released by hydrolyzing the ATP forces the myosin head into a highly strained conformation. The energy from the ATP is stored in the strained myosin head. The strained head can then weakly bind to a subunit of an actin filament. The weak interaction between the actin and strained myosin results in the release of the inorganic phosphate. The strained myosin, with ADP still bound, then binds very tightly to its actin subunit. Once tightly bound, the myosin head undergoes a conformational change that pulls the actin filament. This is the "power stroke", at the end of which ADP is released from the myosin head. The myosin head can then bind another molecule of ATP and release the actin. The free myosin can then hydrolyze the ATP, and the cycle continues. A myosin head may go through about 5 contraction cycles per second.

The key point here is that the hydrolysis of ATP is not directly coupled to the power stroke. Rather, the energy from ATP hydrolysis is stored briefly in a strained protein conformation and subsequently released. This is a theme running through many aspects of cell biology. Ion pumps and multi-enzyme macromolecule synthesizing systems similarly capture energy in strained protein conformations.

In addition to placing muscles into three groups based on structure, flight muscles are grouped into two sorts based upon a functional relationship between nerve firings and muscle contraction. In synchronous flight muscle systems, there is a one-to-one correspondence between each nerve impulse and subsequent muscle contraction. Synchronous flight systems are also called neurogenic systems and include Odonata, Orthoptera and Lepidoptera. In asynchronous flight systems, a single nerve impulse gives rise to several muscle contractions. Asynchronous systems are also known as myogenic systems because the stretch of one set of muscles by their antagonistic pair mates stimulates contraction: the contraction is caused by the muscle, myo, rather than nerve, neuro. Diptera and Hymenoptera are examples of myogenic fliers.

The synchronous flight muscle systems are associated with tubular and close-packed flight muscles, and are characterized by relatively slow wing-beat frequencies. Butterflies produce about 4 to 20 beats/second and locusts about 15 to 20. On the other side, the myogenic systems are associated with fibrillar muscles and quite high wing-beat frequencies. Bees and some flies, for examples, operate at about 190 beats/second, with extremes of up to about 1000 beats/second in some very small flies. The highest frequency recorded is 2218 Hz in the midge Forcips, tested at high temperature and with the wings clipped.

At first inspection, the contractile process as we understand it may not seem consistent with such high frequencies. Let us try to reconcile this. In both neurogenic and myogenic systems the muscles are attached such that a very short contraction permits the full excursion of wing movement. This is more extreme in myogenic systems. For example, in Sarcophaga flight muscles contract by only about 1 per cent of their length during flight.

Myogenic muscles can be made to rhymetically contract if the muscle is stretched and set into oscillation by a mechanical device. Stretching the fibrile produces a slightly delayed contraction, and releasing the tension produces a slightly delayed relaxation. The muscle works by adding energy to maintain an already oscillating system. In the whole animal the thoracic box and wings together form a resonating system, and the timing and duration of muscle contraction is mechanically induced. The operation is started by a direct nervous command; after that first impulse, further nervous input regulates the availability of calcium which in turn regulates the power output of the mechanical system by way of the regulatory proteins we mentioned, that is, native tropomyosin. The fact that the muscle is stimulated to contract by stretching rather than by a nervous impulse gives rise to the term myogenic.

In synchronous fliers, clipping the wings has little effect on frequency because the basic frequency is programmed into the thoracic ganglia. In flies and other myogenic fliers, however, clipping the wings nearly doubles with wing beat frequency without a change in membrane potential spike frequency. These muscles can not be driven by electrical stimulation to go as fast as their normal frequency.

Again, this points to the mechanical nature of the asynchronous flight muscle system. In the whole animal the muscles are not attached to the wings, but to particular elements of the thorax such that the wings are part of a lever with two stable positions - up and down. When muscle contractions move one arm of the lever beyond the mid-point, the skeletal elements spring in, which shortens the muscle and releases tension. But at the same time this motion stretches the pair-mate, causing it to contract. The contraction moves the lever in the other direction, and tension is redeveloped as the lever reaches the mid-point. This oscillating cycle goes on as long as the muscle is maintained in an active state by a relatively small number of nerve impulses.

In the presence of calcium, which is released from the sarcoplasmic reticulum due to nerve impulses, myosin ATPase splits ATP during oscillatory work. Stretch itself increases the activity of calcium-activated myosin ATPase - here stretch acts like an increase in calcium. This is also consistent with differences between myogenic and neurogenic systems. In neurogenic systems, each wing beat may be associated with a contraction-relaxation cycle, with substantial movement of calcium out of and back into the sarcoplasmic reticulum. Asynchronous muscles are less sensitive to calcium flux, which is also correlated with reduced development of sarcoplasmic reticulum.

GAS EXCHANGE DURING FLIGHT

We saw that muscle contraction results from the interaction of thick myosin and thin filaments. Muscle contraction requires large amounts of energy in the form of ATP, which is produced by energy-yielding pathways during flight activity. Most of the ATP is formed within mitochondrial membranes. Mitochondria require large amounts of oxygen to support energy metabolism. The oxygen is joined with hydrogen atoms to form water at the last step in the electron transport chain. Carbon dioxide is formed as pyruvate is decarboxylated before entering the Kreb's cycle, and at two steps within the Kreb's cycle. On a mass-specific basis, some insect flight muscles are capable of the highest metabolic rates of any muscles in the biosphere. These high metabolic rates are supported by efficient gas exchange systems responsible for transporting sufficient amount of oxygen to mitochondria to support flight activity, and transporting carbon dioxide out of the insect.

The transport of oxygen and carbon dioxide into and out of organisms is also called respiration. In insects respiration is conducted by the trachael system. The main anatomical points are the paired spiracular openings leading into a series of tubes. The spiracles often feature elaborate opening and closing mechanisms that are physiologically controlled to regulate air flow and reduce water loss from the insect. The largest number of spiracles is ten pairs: two thoracic and eight abdominal. There are fewer spiracles in some aquatic insects. Some insects have a single pair of spiracles and a few small insect have no functional spiracles. They rely on various form of cutaneous respiration.

The trachael systems of many insect species also include relatively large air sacs. These air sacs can be compressed to force air into and out of the insect body. They sometimes also function in thermoregulation by insulating one body compartment from another.

The trachael system is really a system of tubes. There are spiracles on either side of the body, leading into longitudinal trunks at the sides. There are also major longitudinal trunks located dorsally and ventrally. These large tubes are connected by commissures. The large trunks branch into smaller tracheae (plural of trachea), which further branch into still smaller tracheae, finally extending to all of the metabolizing tissues. The smallest tracheae finally give rise to tracheoles that lead to cell membranes. The branching tubes become progressively smaller as they ramify through the body, the smallest are about 0.1 micron in diameter at the tracheoles. There is no clear functional distinction between tracheae and tracheoles. Tracheoles are formed and remain within cells called tracheoblasts, which are derived from epidermal cells. Hence tracheoles are always intracellular, and thereby differ from tracheae. The trachael tubes are joined in a continuous system and the branching brings the tubes, and an air supply, into contact with nearly all tissues.

While the longitudinal trunks form a more or less continuous system throughout the animal, the tracheal system in various body segments may be isolated from one another. This is usually accomplished by simply funnelling some of the trachael tubes as they ramify through the body. Let us consider the desert locust. Here, the tracheal system in the head is isolated from the rest of the body due to narrow parts of some tracheal tubes, and by degeneration of other tubes. It is thought that this arrangement facilitates a plentiful supply of air to the brain and other cephalic organs. The thorax is similarly isolated from the rest of the body. Even the two sides of the thorax are isolated from one another. Again, this is thought to ensure that plenty of air will get to the flight muscles.

The general pattern of supplying air to flight muscles is similar in all larger insects. Each muscle has a primary supply consisting of a large tracheal trunk or an air sac that lies along the side or through muscles. When tracheae, rather than air sacs run along muscles, the tracheae usually widen into an air sac immediately beyond the muscles. The functional significance of this arrangement probably arises from a point in fluid mechanics: the velocity of the moving air decreases when the air reaches the wide conduit. This allows time for more thorough gas exchange between flight muscles and tracheal tubes. In the close space between tracheae and flight muscle, small, regularly spaced tracheae arise at right angles from the primary supply.

These smaller tubes branch into still smaller tubes and ramify into the flight muscles. As mentioned, tracheoles form the most intimate contact between air supply and muscles. The tracheoles lie along the side of tubular muscles; they indent close-packed and fibrillar muscles. In this arrangement the tracheoles are functionally internal to the muscle while remaining topologically external.

Let us move on to the physiology of gas exchange. Oxygen enters the insect at the spiracles, moves through the tracheal tubes, through the tracheoles, finally arriving at the tissues; once at the tissue, oxygen passes on to the mitochondria, the site of utilization in energy metabolism. Carbon dioxide, in its turn, follows this path in the reverse direction. There are two phases of gas transport. One is an air tube transport phase and the other is a tissue transport phase. The later occurs in the aqueous melieu of the cells.

In the early days of insect physiology, it was not at all clear from a simple consideration of the tracheal system that diffusion of gasses could account for the movement of oxygen and carbon dioxide to and from the tissues. Although diffusion was suggested as long ago as 1816, many other ideas were considered before the diffusion theory was finally accepted.

Some of the factors that impact on the rate of diffusion of gas can be appreciated from a mathematical expression known as Fick's First Law of Diffusion:

$$dv/dt = -ADdx/dt$$

This expresses the idea that the amount of gas diffusion in a given time is related to the product of the area (A) times the diffusion coefficient of a particular gas (D) times the concentration gradient over the distance x. In the early years of this century Krogh considered these factors, plus the length and diameter of tracheae, plus the oxygen utilization rates of insects to calculate what difference in oxygen pressure between the atmosphere and the tracheal endings - that is, the concentration gradient - would be necessary to maintain the supply of oxygen that is actually consumed by insects. His calculations indicated that the oxygen pressure at the tissue level needed to be only 2 or 3 per cent below atmospheric oxygen pressure. In some dragonflies, the oxygen pressure at tissue level works out to be about 5 per cent below atmospheric levels. This was the argument that led to the general acceptance of the diffusion theory in tracheal transport of oxygen. As for the reverse path, the carbon dioxide pressure in the atmosphere is nearly negligible, and there is always an adequate gradient for the outward diffusion of this gas.

Recall that there are two diffusion phases, an air tube phase and a tissue phase. Krogh also calculated permeability constants for oxygen. Permeability constants are a linear function of diffusion coefficients. Here are the constants:

AIR 11.0 = $cm^2/min/atm$

WATER = 0.000034 $cm^2/min/atm$

MUSCLE = 0.000014 $cm^2/min/atm$

Inspection of these values reveals that oxygen is more permeable in air than it is in water or in tissues. We can infer that oxygen diffuse faster in air than in tissue. To be more precise, oxygen diffuses 785,714.28571 times faster in air than in tissue. For a round number that we can remember, it is on the order of 1 million times faster.

Consider a concentration gradient along a tracheal tube of 3.0 cm length. If the gradient were sufficient for the diffusion of oxygen over the 3 cm of air phase diffusion, that same gradient would be adequate for a diffusion distance of about 0.000003818 cm in the tissue phase. This is about 0.04 microns. These calculations emphasize the point that tracheoles must be quite close to mitochondria for diffusion in the tissue phase to provide adequate oxygen transport. The late Weis-Fogh, an insect scientist at Cambridge University, predicted that in dragon fly tubular muscles the tracheoles would have to be less than about 10 microns apart. The tracheoles in fibrillar muscles are about 3 - 5 microns apart.

In resting and in some flying insects diffusion through the air tube and the tissue phases will supply enough oxygen to the flight muscles to support energy metabolism. However, in larger insects, diffusion alone will not supply enough oxygen to support flight activity. Various insect groups have evolved mechanisms of ventilating the major trunks of the tracheal system during times of high metabolic activity. Recall the expression $dv/dt = -AD\ dx/dt$, showing us that the concentration gradient if a function of the distance of the diffusion path. As the diffusion distance gets longer, the concentration gradient becomes smaller and the diffusion rate or flow decreases accordingly. The situation is turned about face by ventilation: the diffusion distance is reduced, the gradient becomes larger and gas exchange is carried out by the same mechanism. Diffusion.

Let us turn, now, to a look at how insects ventilate the major trunks of the tracheal system during flight. One mechanism of ventilation is called draft ventilation. This is seen in cerambycid, or long-horn beetles. We have seen that the spiracles in most insects are located on either side of the thorax and abdomen. The situation is slightly modified in beetles, which feature larger spiracles located at the front of the thorax. These spiracles are open during flight. Air enters through the second pair of spiracles and exits through the third pair, thereby creating a unidirectional, ram-jet flow. These are sort of turbo charged insects.

Another ventilation action is called thoracic pumping. As the muscles of the thorax contract during the wing stroke, the thoracic box alternately increases and decreases in volume. This action presses on air sacs in the thorax, causing them to experience volume changes during each wing stroke. These volume changes force air into and out of the thoracic air sacs, thereby ventilating them as a consequence of normal wing movements.

Many insects actively ventilate the thorax by compressing large abdominal air sacs. This is known as abdominal pumping, and is done by contracting intersegmental muscles in the abdomen. Abdominal pumping causes increased hydrostatic pressure and organ movements which in turn press on the air sacs. Air is forced out of the sacs. Air is rhymthmically drawn into the air sacs by muscular or elastic expansion of the abdomen. True bugs and beetles move the turgum up and down. Dipterans and hymenopterans move both the turgum and sternum. Some hymenopterans can also telescope the abdomen in and out. In all of these ventilation mechanisms, it is the major tracheal trunks that are ventilated. Diffusion is still the driving force for gas transport within the secondary tracheae and tissues.

We can identify the sites of gas exchange in animals. The tracheae and tracheoles are respiratory surfaces in insects, in the same sense that lungs are the respiratory surfaces in mammals and gills are the respiratory surfaces in fish. Because water readily traverses these surfaces, respiratory surfaces are major avenues of water loss in terrestrial organisms. Water loss is reduced when the respiratory surfaces are somehow covered. Internalization of the respiratory surfaces is regarded as a major adaptation for terrestrial life. Whereas aquatic animals often have external gills, water loss rates across these surfaces would be far too high in terrestrial habitats.

Although their respiratory surfaces are internalized, insects rapidly desiccate when their spiracles are open. Many insects regulate spiracle opening in response to respiratory requirements. This is best studied in a large insect, the desert locust, *Schistocerca gregaria*. There are two modes of ventilation in this grasshopper, one when the insect is at rest, and another during flight. During rest spiracles 1 and 4 on both sides open during abdominal expansion, which results in air inspiration. Then spiracles 1 and 4 close, and spiracle pair 10 opens while the abdomen contracts, and air is expelled. These actions create a slow, unidirectional air flow of about 30 ml/ gram body weight/ hour.

Spiracles 2 and 3 are open continually during flight. Thoracic pumping causes alternate inspiration and expiration through these openings, at about 250 ml/ gram body weight/ hour. Spiracle pairs 1 and 4 through 10 may open and close rhythmically, but the precise pattern is uncertain. Abdominal pumping moves about 180 ml of air in and out of the abdomen, and about 30 ml/ gram body weight/ hour is moved unidirectionally from the thorax and out of the abdomen. The bulk of air movement is tidal.

Let us summarize aspects of respiration in insects by looking at the features of an effective respiratory surface:

1. *Moisture*: Tracheoles are buried in tissues and they are permanently moist.
2. *Large Surface Area*: Tracheoles are the respiratory surface. They are about 300 to 400 microns long and they occur every 3 to 5 microns. You can calculate the approximate surface area by assuming the

tracheoles are cylinders of about 0.1 micron diameter. The circumference of a circle is calculated as Pi times diameter, and area of the cylinder is calculated as circumference times length of the cylinder: (Pi)(0.1 microns)(400 microns) = 125.7 microns2 of surface area per tracheole. Given a very large number of tracheoles, we can calculate a substantial respiratory surface area.

3. *Permeability to Gases*: Tracheoles are openly permeable to gases.

The respiratory systems of insects are adaptive in ways that go beyond the physiology of gas exchange. One point is the system is a series of tubes and air sacs. The tubes are relatively light, and this lowers the specific gravity of insect bodies. The lower specific gravity is important in aerodynamic aspects of flight, and also provides a degree of buoyancy in aquatic insects. In the larvae of some dipterans, the tracheae enable the buoyancy to be adjusted.

Another point relates to growth. An important feature of insects is that they can not change their overall size within the span of any one instar. Yet growth must occur. Air sacs, which are collapsible, allow for the growth of organs within the body. For example, at the beginning of a juvenile instar in *Locusta,* the tracheal system occupies about 42 per cent of the body volume. By the end of an instar, the tracheal system may be reduced to about 4 per cent of the body volume, due to growth of organs.

A third point also relates to body growth. Many insects appear to expand the tracheal system immediately after a molt, before new cuticle is completed. This is thought to help increase body size before the cuticle hardens in the following instar.

Air sacs are also involved in thermoregulation in some insects. In the dragonfly *Aeschna* air sacs in the thorax insulate the thorax from the abdomen. Similarly, abdominal air sacs in bumblebees thermally insulate the abdomen from the thorax. This is important in selective temperature regulation of the thorax during flight and of the abdomen during brood incubation.

Chapter 2

Insects Anatomy

NERVOUS SYSTEMS

All organisms need to communicate with the biotic and abiotic components of their environments. Organisms also communicate within themselves. We saw during our discussions of homeostatic mechanisms and physiological adaptations that insects need to register conditions and changes within themselves and in their environments, and then respond to the conditions. The physiological basis of homeostatic mechanisms is that an organism can somehow measure aspects of the parameters that are regulated. For examples, changes in hemolymph salt concentrations and thoracic temperature need to be registered before corrective action can be initiated. Insects also respond to changes in their environments. Mechanisms of entering diapause seem a particularly good example. Here, insects register token stimuli, most often changes in photoperiod.

These token stimuli are translated in neuroendocrine events, which in turn are translated into physiological changes. Nervous systems play crucial roles in perceiving environmental conditions, integrating environmental signals into meaningful information, and then producing responses in the forms of internal, physiological events. In another, speedier set of conditions, try imagining how an insect would fly if it could not see where it was going, or could not determine orientation with respect to its horizon. Within a flying insect, again, try imagining losing control of the small, direct flight muscles that control pronation of wings, or loss of coordination of the major flight muscles. Also consider various sorts of homeostasis during flight: internal temperatures are measured, then hemolymph flows adjusted accordingly; available fuels in hemolymph are oxidized, then replaced from stores quite distant from flight muscles. It would be difficult to regard any area of insect biology that does not depend importantly upon nervous systems.

Understanding insect nervous systems also has a certain practical significance because most insecticides work by poisoning one or another area of nerve function.

The nervous system is made of neurons. Every tissue is composed of cells. Fat bodies are mostly made of trophocyctes. Muscles are composed of muscle cells. Similarly, nervous systems are composed of nerve cells, which are also called neurons. The first figure shows the general structures of nerve cells. Nerve cells have cell bodies, called perikarya (pleural of parikaron). Most major organelles, including cell nuclei, endoplasmic reticula, Golgi bodies, mitochondria and so forth are found in perikarya. Long projections known as axons (pleural of axon) extend from cell bodies; axons carry nerve impulses to their ends, often called terminal arborizations. These arborizations are typically juxtaposed very closely to other nerves or to muscles. At these points, called synapses, electrical impulses are transduced into chemical signals, which are received at membranes of other nerves or muscles. Nerve cells receive stimuli at their dendrites (pleural of dendrite). Dendrites can be found directly on perikarya or at ends of axons.

We consider three kinds of neurons. The upper drawing shows a monopolar neuron, which have a single projection from parakarya. This single projection branches so that there is a dendrite at one end and terminal aborization at the other. Most insect neurons are monopolar. The middle sketch shows a bipolar neuron; these neurons are commonly associated with sense cells, and feature a dendrite at a distal end and a terminal aborization at the other. This arrangement is suited for direct transmission of environmental stimuli to the central nervous system. Multipolar neurons have several axons projecting from perikarya; this cell type occurs in some ganglia, and is also connected with stretch receptors.

Information from outside of an insect, such as light or mechanical stimulus, is most likely conducted from peripheral sense cells directly to central nervous system ganglia by axons without synaptic interruptions. These axons are called afferent axons. Other axons are involved in conducting motor commands from central nervous system ganglia, wherein lie the cell bodies, to muscles; these axons are called efferent or motor axons. Most afferent neurons synapse with other neurons that in turn synapse with still other neurons or else with efferent neurons. The neurons that form connections between afferent and efferent neurons are called interneurones. Nerve fibres are generally aggregates of several or many neurons.

THE CENTRAL NERVOUS SYSTEM IS NOT CENTRALLY LOCATED

Unlike vertebrates, in which a single, relatively large brain is encased in a skull, central nervous systems of insects are composed of variously distributed ganglia. The major association centre is called the brain. The other important ganglia are subesophageal ganglion, frontal ganglion, up to three thoracic ganglia and up to eight abdominal ganglia.

Three major regions of brains are recognized: protocerebrum, deutocerebrum and tritocerebrum. The protocerebrum is the most complex part of insect brains; it has two lobes, and is continuous with the optic lobes. The deutocerebrum contains antennal lobes, from which project both afferent and efferent neurons. The tritocerebrum is a small, bilobed ganglion. Major nerve tracts connect the tritocerebrum to other ganglia. The tritocerebrum is linked to subesophageal ganglion by circumesophageal connectives and to the frontal ganglion by frontal commisure. This part of the brain has efferent and afferent connectives to certain mouthparts.

The subesophageal ganglion is the first ganglion in the ventral nerve cord. This ganglion has motor and sensory connections to salivary glands, mouthparts, and neck. Most insects have three thoracic ganglia, each with connectives to muscles and sensilla. Abdominal ganglia have connections to muscles, though usually fewer than thoracic ganglia.

Numbers of ganglia vary among groups of insects. The thoracic and abdominal ganglia are fused into a single compound ganglion in houseflies. This is an example of extreme fusion. The least fusion is shown in a thysanuran. The last abdominal ganglion is always a compound structure, derived from the ganglia of the last four segments.

The next figure sets forth the structure of a ganglion. The central nervous system is completely invested in what is called the nerve sheath, the most outer part of which is a non-cellular neural lamella and the inner part of which is called perineurium, composed of a single layer of cells. The perineurium is thought to produce and secrete the mucoprotein and mucopolysaccharide neural lamella. The perineurium serves as a blood-brain barrier, and as such it regulates the chemical environment of nerve cells and transports nutrients and other materials between hemolymph and nerve cells. The centres of ganglia are made up of axons and fibres of afferent, efferent and interneurones. These centres specifically do not contain cell bodies, and they are called neuropile. Within neuropiles, some axons and fibres are orientated such that they form fibre tracts. Cell bodies, or perikarya, are found around the periphery of ganglia.

The next figure shows that inter ganglionic connectives are similarly arranged, with neural lamella and perineurium forming a sheath around the tracts and giant fibres running along in the middle. Within neuropiles and connectives, each neuron is surrounded by glial cells which form a protective barrier around nerve cells. Glial cells can have a number of arrangements, and are often folded around axons, in which case they are called mesaxon. Glial cells probably insulate nerve processes from each other to reduce uncontrolled electrical noise or "cross-talk" from one nerve to another. Glial folds do not occur around synapses.

PHYSIOLOGY OF NERVE CELLS

An electric current is defined as the amount of charge passing through a conductor in some unit of time. In metallic electrical wires charge is carried by electrons, but in physiological electrolytic solutions current can only flow by the movement of ions because there are no free electrons. Movement of charged particles requires work be done, which requires expenditure of energy. Current will only flow through a conductor if there is a potential difference across it. A potential difference is the amount of work required to move a single charge from one place to another. For our purposes, the electrical events associated with nerve cells relate to movements of ions through nerve cell membranes.

These ion movements are also electrical currents. The major ions in nerve physiology are sodium (Na) which is generally in higher concentration outside of axons, potassium (K) which is generally in higher concentration inside, and chloride (Cl). The important point here is that these ions have different permeabilities. These different permeabilities are usually spoken of as different conductances. In electrophysiology we think of an electrical current (movement of ions) in terms of the potential difference that drives the current and of the conductance of the driven ion. A sodium current is written like this:

$$I_{Na} = (V_{Na}) \times (g_{Na}),$$

where V is potential difference and g is conductance. We recall from our discussions of osmoregulation that have talked about cells in various sorts of solutions. We saw that ions tend to move down their concentration gradients, from high to low concentrations. What we want to introduce now is one more feature of concentration gradients. When determining membrane equilibrium we need to consider both chemical gradient and electrical potential.

Let us imagine a two chambers separated by a membrane. If there is a high Na concentration on one side and a low Na concentration on the other, there will be a greater flux of Na in the down-hill direction, compared to the up-hill direction. This translates into a net down-hill Na flux. The Na flux would generate a net current flow. Now if we apply a potential difference between the two chambers, the potential difference can be twiddled such that the chemical concentration difference can be balanced by the potential difference. In theory an appropriate potential difference can be used to offset any chemical concentration difference. We can express this in slight more formal terms.

Nernst formulated a mathematical expression to relate a potential difference to a given concentration difference:

$$E + -(RT/zF) \times \log_e [c]_a/[c]_b$$

where R is the gas constant, T absolute temperature, C concentration, z valance of an ion and F the Faraday. All the constants can be treated accordingly, and for 37°C we can write:

$$E = -60 \log (C)/(C')$$

Here E is potential difference in millivolts and this is called the electrochemical equilibrium potential, or the Nernst equation. This is extremely useful for electrophysiologists because concentrations of ions can be measured and potential differences across membranes can be measured. Hence, electrophysiological events can be expressed in measurable terms. This becomes the basis for research on ionic mechanisms of nerve conduction.

The last complicating point is that resting potentials of nerve cells are made up from interactions of three ions, so the Nernst equation need be re-written to account for that:

$$V = -RT/F \times \log_e g_K[k^+]_I = g_{Na} [Na^+]_I + g_{Cl} [Cl^-]_I$$

$$g_K[k^+]_O = g_{Na} [Na^+]_O + g_{Cl} [Cl^-]_O$$

where I stands for intracellular concentration and O stands for extracellular (outside) concentration. This is the Goldmann field equation, and for the giant squid axon, this model predicts the resting potential of membranes fairly well.

The physiology of maintaining a potential difference across membranes takes us back to the physiology of moving ions across membranes. One of the most well studied ion pumps is the sodium-potassium ATPase. In this system three Na ions are moved outwardly while two K ions are moved inwardly at the expense of 1 ATP. Hence, this pump requires extracellular K to function. Ions typically leak across membranes, and ion pumps are probably running constantly to maintain resting potentials in nerve cells.

Now let us change gears slightly and look at examples of nerve physiology in action. The next figure shows examples of what are called spike frequencies. The first example shows spiking of antagonistic motor neurons in a walking insect. We see a burst of action potentials, then a relatively quiet period, then another burst of action potentials. The second example shows responses of olfactory receptor neurons of a male moth to different levels of a sex attractant pheromone, tetradecenyl acetate. Again, at high levels of sex pheromone, we see very high spike frequencies.

These are examples of spiking in efferent and afferent neurons. Each spike represents an action potential. The next figure shows an action potential resolved at sub-millisecond level, along with a diagram of the ion dynamics associated with the action potential. Again, the resting potential is maintained by ion pumps. About -70 mV is a common sort of resting potential. The original ion distribution. As an impulse moves along an axon, the first phase of an action potential is caused by a sharp increase in Na conductance.

This is due to opening of voltage-sensitive ion channels, so that Na ions can follow their concentration gradients across the axon membrane into the axon. A rapid change in potential difference across the membrane follows, perhaps as much as

100 mV. Na conductance is high for about 0.5 millisecond, whereupon K conductance rises, and K ions rush out of the axon. The sharp increase in outward K flux causes the falling phase of the spike potential. The high K conductance causes an overshoot, known as the positive phase. The negative after potential results from a local high accumulation of K outside the axon membrane. Na-K ATPase eventually brings all ions back to resting potential concentrations.

A few points remain to be addressed. First, a very small number of ions is involved in an action potential, on the order of 100 ions, or less than one millionth of available K ions inside an axon. Second, action potentials usually move in only one direction due to refractoriness. This can be explained by considering conductances. Na conductance is very much higher than K conductance. Third, impulses must move along axons for information to flow. So we want to consider how impulses move along an axon. The local circuit theory is based on the idea that areas of an axon next to a point of action potential are negatively charged so that a local current flows away from a point of action potential on the inside, and towards it on the outside.

TRANSMISSION ACROSS SYNAPSES IS CHEMICAL, NOT ELECTRICAL

The important features include presynaptic vesicles, many filled with transmitter substance, usually acetylcholine in nerve-nerve synapses, and L-glutamate in nerve-muscle junctions. Gamma-aminobutyric acid (GABA) is the usual transmitter chemical in inhibitory junctions. Other features are presynaptic membrane, and postsynaptic membranes. Finally two proteins are important, acetylcholine esterase and acetylcholine receptor sites. This drawing shows the synaptic cleft, which separates pre- and postsynaptic membranes by 200 to 500 angstrom units.

Contemporary wisdom holds that depolarization of presynaptic membrane leads to fusion of presynaptic vesicles with membrane, and release of acetylcholine by a process of exocytosis. Vesicles contain on the order of a few thousand molecules of transmitter; each depolarization causes about 100 or so vesicles fuse with presynaptic membranes and release their molecules. Acetylcholine molecules move randomly as in free solution within synaptic clefts, and it is about equally possible for any given molecule to meet acetylcholine esterase or acetylcholine receptor sites. The enzyme hydrolyzes acetylcholine to acetate and choline. The acetylcholine molecules that bind to their receptor sites eventually cause a depolarization of the postsynaptic membrane called excitatory postsynaptic potential (EPSP), which is transmitted to the axon. Axon depolarizations and continuation of the nerve impulse follow.

INSECT VISION

Vision is the perception of light. Insects perceive light through three classes of sensory organs. Most adult insects have a single pair of compound eyes. Larvae of hemimetabolous insects and most adults have usually three simple eyes called ocelli. These are typically located on the dorsum of the head capsule, and they are sometimes called dorsal ocelli. Larvae of holometabolous insects do not have compound eyes. These insects perceive forms, to a limited extent through stemmata located on either side of the head.

COMPOUND EYES

The compound eyes are prominent features of adult insects. The eyes occupy a fairly large portion of the surface of insect heads, and they facilitate a rather wide field of vision. Compounds eyes are not present in all insects. They are reduced or absent in parasitic forms, many soil insects, and in some species that live in very dark places, such as caves.

The basic unit of compound eyes is the ommatidium (singular of ommatidia). Ommatidia vary in size and number among insect groups. At one extreme, dragonflies have thousands of ommatidia in each compound eye; many insects have far fewer, and at the other extreme, the workers of the ant species *Pomera punctatissima* have only one ommatidium in each eye. The sizes of ommatidia vary from about 5 to 40 microns in diameter. Sizes vary among species and even within a single compound eye. In some dragonflies, for example, the dorsal units are considerably larger than the ventral ones.

A figure shows the structure of a single ommatidium. The ommatidium rests on a basement membrane. The height of the eye comes from the long retinula cells and secondary pigment cells. The corneagen cells (sometimes called primary pigment cells) rests atop the long retinula and secondary pigment cells. The crystalline cone lies within the corneagen cells. The surface of the ommatidium is covered with the corneal lens. The corneal lens is a specialized part of insect cuticle. It is secreted by the corneagen cells, which are specialized epidermal cells. The crystalline cone is a clear, intracellular structure synthesized within Sempter cells. The primary pigment cells surround the Sempter cells. Not all compounds eyes feature a crystalline cone; those that do are called eucone eyes.

The corneal lens and crystalline cone make up the optical part of ommatidia. The sensory elements are just under the optical apparatus. The retinula cells are long neurons. Part of each retinula cell is a specialized area known as a rhabdomere. The rhabdom indicated in the figure is a collection of up to eight rhabdomeres. A nerve axon from each retinula cell projects through the basement membrane into the optic nerve. Ommatidia are functionally isolated because the retinula cells are surrounded by the secondary pigment cells.

Compound eyes are commonly categorized as either apposition eyes or superposition eyes. There is no separation between the corneal layer and the photoreceptors in apposition eyes. There is a clear space between the two units in superpositions eyes. Some authors use the expression "clear zone eyes" for the superposition eyes. Superposition eyes are generally found in crepuscular and nocturnal insects, while apposition eyes occur in diurnal insects.

The crystalline cone couples the lens and photoreceptors in apposition eyes. Some higher Diptera have pseudocone eyes. Crystalline cones do not occur in pseudocone eyes, and optical coupling is by means of a gelatinous substance that is contained in a two-celled structure. Still another arrangement is found in some apposition eyes that lack solid cones or gelatinous pseudocones. In these acone eyes four flat transparent cells are found in place of the cones. In superposition eyes the space between the corneal lines and the photoreceptors is traversed by crystalline tracts. These are thought to act as wave-guides that direct light to the photoreceptors.

The eyes of some insects feature a tracheal tapetum, as shown in a figure. Tracheal tapeta are known from several Lepidoptera. Small tracheae run parallel with the longer cells of the ommatidia. It has been suggested that these structures function as interference reflector filters.

Light sensitivity of insect eyes varies according to the state of light- adaptation or dark-adaptation of the eyes. In bright light, the eye is less sensitive to light, and the eye is regarded as light-adapted. Maximum sensitivity occurs in darker conditions, when the eye is fully dark-adapted. Some insect eyes can change light sensitivity by about three orders of magnitude within a few minutes of changing light conditions.

There are a number of mechanisms of adaptation to light and dark conditions. One mechanism relates to the biochemistry of vision. The visual pigment is broken down by interaction with light. In daylight the breakdown rate can equal or exceed the replacement rate. This leads to decreased light sensitivity, and partly explains light adaptation. In the dark, the visual pigment accumulates, and the insect becomes dark adapted.

A second mechanism of adaptation is the movement of screening pigments. Screening pigments are located in the retinula cells. In light-adapted eyes granules of screening pigments surround the rhabdom. These pigments absorb light, and they have the effect of optically isolating individual ommatidia. The endoplasmic reticulum of dark-adapted eyes forms large, clear vesicles. This forms a clear space around the rhabdom, so that a greater amount of light falls upon the rhabdom.

Vision involves the transduction of light energy into a bioelectric signal within the nervous system. The first events in this process take place in the retinula cells. The fine structure of rhabdomeres consists of thousands of closely packed tubules. These tubules are about 500 A in diameter and one micron long. They are aligned at right angles to the long axis of the rhabdom. The visual pigments occur mainly in these rhabdomeric microvilli. It has been suggested that the small diameter of each microvillus inhibits free rotation of visual pigments. This specific orientation may be the molecular basis of insects' sensitivity to polarized light.

Photobiological processes occur in a narrow band of the electromagnetic spectrum between 300 and 700 nm. Photons in this region of the spectrum have enough energy for photochemical interactions, but not enough energy to disrupt macromolecules. Visual pigments initiate vision by absorbing light in this region. These pigments are a class of membrane bound proteins known as opsins that are conjugated with a chromophore. Visual pigments whose chromophore is retinal are called rhodopsins. The visual pigments of all invertebrates, including insects, crustaceans and squids, are all rhodopsins.

There are a number of geometric isomers of retinal. The 11-cis isomer is the chromophore of native rhodopsin. Free retinal absorbs light strongly at about 380 nm. When 11-cis-retinal is linked to opsin to form rhodopsin, the main absorption peak is shifted. The maximum absorption values of different rhodopsins range from about 345 nm (ultraviolet) to as high as 610 nm (red).

The absorbing properties of any given rhodopsin is related to the disposition of charged opsin groups around the chromophore. These interactions between the side-groups of the amino acids within the opsins and the chromophores are thought to modify the p electron orbital of the chromophores, and thereby modify the absorbance spectra of rhodopsins.

When the chromophore of rhodopsin absorbs a photon, the chromophore undergoes an isomerization from the 11-cis to the all-trans configuration. The rhodopsins of insects and other invertebrates differ from vertebrate rhodopsins on the point of the chemical events that follow the isomerization step. Vertebrate rhodopsins go through a series of spectrally distinct intermediates. The process is called bleaching because the absorption shifts from visible to ultraviolet wavelengths, which we can not see. The vertebrate process can lead to hydrolysis of the opsin and chromophore. Invertebrate rhodopsins generally do not bleach. Light transforms the rhodopsins to a stable intermediate. The isomerized 11-cis chromophore stays in place in the microvilli. The stable intermediates have spectral properties similar to metarhodopsin I of vertebrates, which has earned them the term metarhodopsins.

The mechanism of transducing light-induced chemical changes in rhodopsins into electrical signals remains the major unresolved issue in understanding the physiology of light receptors. Several milliseconds are required before the absorption of light can be recorded as electrical activity in the receptor cell membrane. Certainly, the phototransduction event occurs during the transition from rhodopsin to metarhodopsin. We can not say much more about the detailed mechanisms of transducing light into bioelectrical signals.

A key requirement in the chemistry of vision is that metarhodopsin, or the bleached pigment of vertebrates, must be reconverted to rhodopsins. In the eyes of vertebrates, this regeneration is the product of a complex biochemical pathway. The pathway involves isomerization back to the 11-cis configuration and assembly with the opsins. Again, invertebrates differ from vertebrates on this aspect of photochemistry. Light is absorbed by rhodopsin and metarhodopsin in invertebrates. Absorption of light transforms metarhodopsin back to rhodopsin. The pigments exist in two stable conformations, and light causes them to switch back and fro from one state to the other. This regeneration is essential because conformational changes in rhodopsin, but not in metarhopsin, leads to phototransduction.

Experiments on insect behaviour indicate that many insect species can distinguish colours. Colour vision depends on the ability to discriminate between light of different wavelengths. Such discrimination is possible because compound eyes have photopigments with differing spectral properties. Many insects have two rhodopsins, one with maximum absorption in ultraviolet wavelengths and one with maximum absorption in green wavelengths. Some species have a third pigment that absorbs maximally in the blue region. A few Lepidoptera have four visual pigments.

DORSAL OCELLI

Dorsal ocelli occur in larvae of hemimetabolous insects and in nearly all adults. Here is the structure of a dorsal ocellus. Although there are variations in structure, a typical ocellus has a single lens that is usually rather thickened. Most ocelli feature a large number, often hundreds, of retinula cells. Rhabdomeres of several cells combine to form a number of rhabdoms. Axons project from the retinula cells through the basement membrane and they end in a synaptic plexus behind the eye. Light causes a sustained depolarization of the retinula cells, however, the biological roles of ocelli remain unknown. It is generally agreed that ocelli allow only poor, if any, perception of form. In some orthopterans the ocelli are active in orientation to a light source.

STEMMATA

The larvae of holometabolous insects have stemmata on the sides of the head capsule. Many larvae have a single stemma (singular of stemmata) on either side, but the number can be as many as six on a side. Here is a sketch of a section through a stemma. Each stemma has a cuticular lens and a crystalline cone distal to the sense cells. Stemmata generally do not produce clear images, but most caterpillars can discriminate some shapes and they can orientate themselves with respect to boundaries.

BIOCHEMISTRY OF INSECT BODY

HOMEOSTATIC MECHANISM

Homeostatic mechanisms maintains stable conditions with respect to any given physiological parameter. Consider homeostasis of thoracic temperature during flight; of hemolymph energy metabolite concentrations; of water balance in terrestrial insects; of osmotic balance; of symbiotic organisms. Two aspects of physiology are implied in discussions of homeostatic mechanisms: first, organisms can register changes in an homeostatic parameter, and second, features of the parameter can be adjusted. In homeostasis of thoracic temperature, it is clear that thoracic temperature is monitored during flight (even if we do not understand the measuring system), and it is clear hemolymph circulation patterns are adjusted in response to changes in thoracic temperature. I introduce a discussion of the insect fat body in terms of homeostasis for a couple of reasons. First, fat body is involved in many homeostatic mechanisms; second, I want to emphasize the point that fat body is an active centre of metabolic activity.

In most insects the fat body serves as a storage depot for food reserves, and sometimes for storage excretion. Lipid reserves are often accumulated in massive quantities in this organ, so that it looks like a body, or organ, of fat. The term fat body was coined during early days of studying insect morphology, and it derives from the fatty appearance of the tissue. This term is deeply entrenched in the literature of our field, so much so that it would be a futile exercise to try working toward a more descriptive expression for this organ. The alternative is to expand our conception of what the term fat body represents.

In addition to its important roles as a storage depot, the fat body of insects functions as a key centre of metabolism and biochemistry. As a metabolic and biochemical centre, the biological significance of fat body is its ability to maintain a balance between resources and requirements during the many phases of an insect's life. On one side of the balance, during times of feeding fat bodies biosynthesize and accumulate not only lipid reserves, but also carbohydrates, amino acids, proteins and other metabolites.

On the other side, fat bodies respond to physiological and biochemical needs in a number of ways, including very high rates of protein biosynthesis, formation and release of trehalose, release of lipids, detoxification of nitrogenous waste products, and biosynthesis of hormones. Many of the responses to physiological needs occur on a relatively large scale, and they can have substantial impact on insect biology. Very often fat body metabolic functions are regulated by various endocrinological mechanisms.

STRUCTURES OF FAT BODIES

The morphology of fat bodies varies considerably in various insect groups. In many species it is a loose aggregation of cells that may be unevenly distributed among all body segments, including the brain case. Fat body cells often surround the major organ systems in insects, including reproductive tissues, alimentary canals, thoracic muscles, and elements of central nervous systems. This close association of fat body cells and other tissues may facilitate exchanges of macromolecules. Fat bodies may also protect organs from physical damage by providing packing. In other species fat bodies are compact cell masses invested in a membranous covering such that they easily can be removed from the insect intact.

Structures of fat bodies may vary according to life stage, sex and reproductive status. Fat bodies can have different functions between sexes, and functions also can differ among life stages. In some species they undergo histological breakdown and reformation during papal stages, though in most species they do not appear to do so.

Fat bodies of all insects are in intimate contact with circulating hemolymph, which is consistent with movements of molecules between the two compartments. The vast majority of the cells in fat bodies are called trophocytes, and it is in these cells that most fat body functions are carried out. Trophocytes are very similar to some hemocytes; the two cell types may be quite closely related. In the homopteran Aleyrodes, for example, fat body cells float individually in hemolymph and can not be clearly distinguished from hemocytes.

Other fat body cell types include urate cells. These cells are associated with storage excretion, in which uric acid crystals are accumulated rather than excreted. Fat bodies of most insect species do not include urate cells. For the sake of complexity, uric acid is stored in trophocytes of some species. Fat bodies of many species also contain mycetocytes, which are involved in housing symbiotic micro-organisms. Many of these symbiants play crucial roles in insect nutrition by providing certain vitamins and other essential nutrients to insects that are not able to acquire these nutrients in the diet.

PROTEIN BIOSYNTHESIS IN FAT BODY

Many of the proteins that are crucial in the lives of insects are biosynthesized in fat bodies. Protein synthesis is often under control of juvenile hormone (JH). One of the most well-studied examples of endocrine regulated protein synthesis in fat bodies is the formation of vitellogenins by adult females. Vitellogenins are the yolk proteins that are accumulated during egg maturation. Fat bodies biosynthesize many other proteins as well, including virtually all of the major proteins found in the hemolymph of larval insects. There are also many examples of specialized proteins in insects. Larvae of the midge *Chironomus* biosynthesize hemoglobins, heme-containing proteins important in storage of oxygen. Hemoglobins can make up more than 80 per cent of the hemolymph proteins in these larvae. The lipoproteins that function in hemolymph transport of diacylglycerol are synthesized in fat bodies. There are many other examples, all bearing on the point that protein synthesis in fat bodies is a central issue in insect biology.

We review the biochemistry of protein synthesis by way of appreciating the importance of this topic. First we recall the expression originally put forth by Francis Crick, and later articulated by Charlie Brown as the central dogma of molecular biology: DNA Makes RNA Makes Protein. As an aside, a group of my biochemical colleagues at the University of Nevada, Reno and myself once entered a biochemical T-shirt logo contest. We added the expression makes lipids to the expression of the central dogma of molecular biology. We did not win. This figure shows that genes are transcribed from DNA to RNA within a cell nucleus.

The transcription process produces a special form of RNA called messenger RNA, or mRNA. Messengers go places, and in this case mRNA makes its way from the nucleus to the cytosol where it joins with ribosomes, modeled in the figure. We want to see that the mRNA is joined with two ribosomal subunits, a small and a large subunit. The next figure shows this is a little more detail. The flow of information is from the nucleus to the cytosol This is followed by a series of steps that include translation of the genetic information borne on the mRNA strand into a polypeptide chain, or string of amino acids. The key here is that each of the amino acids can be attached to a specific RNA molecule known as transfer RNA, or tRNA. One portion of tRNA molecules feature an anticodon that matches up to codons on the mRNA strand.

Ribosomes have sites for two tRNA molecules, one called a peptidyl site and the other called an aminoacyl site. While two tRNA molecules are juxtaposed within the ribosome a peptide bond is formed between the amino acids attached to each tRNA. At this point the tRNA molecule in the peptidyl site is released and the ribosome scoots one step along the mRNA strand, thereby shifting the tRNA molecule in the aminoacyl site to the peptidyl site and opening a space for a new tRNA molecule. The cycle goes on with addition of the new tRNA molecule, formation of another peptide bond, release of a free tRNA molecule and another ribosomal shift. A copy of the universal genetic code is appended to the end of this column. Please note, for example, that the AGG in the mRNA strand shown above specifies the amino acid arginine.

Protein synthesis begins with transcription of a DNA strand within a cell nucleus. JH expresses its effects on protein synthesis within fat body cells by interacting with DNA. JH receptor proteins are found within the cytosol of cells. JH forms a complex with its specific receptor, then the hormone-receptor complex moves into the nucleus. It is the JH-receptor complex that interacts with DNA.

CARBOHYDRATE METABOLISM IN FAT BODY

The natural foods of many insects include substantial amounts of carbohydrates, or sugars, either in the form of simple sugars or as polymers. Following absorption across midgut epithelium, sugars are removed from circulating hemolymph into fat bodies. Sugars are assembled into large macromolecules of glycogen and stored within fat bodies (try to imagine a discussion on the physiology of the sugar body).

The polymerization of simple sugars into glycogen is called glycogen synthesis, and is carried out enzymatically by glycogen synthetase. It has been suggested that glycogen synthetase may be under control of a sort of hypoglycemic hormone. There are a number of lines of evidence for factors that reduce hemolymph sugar levels, but the specific points of control remain unknown.

But there is also evidence for endocrine regulation of carbohydrate mobilization. Various names have been applied to hormones that increase hemolymph sugar levels, including hypertrehalosemic factor and glucogon-like hormone. Most recently people have been using the term trehalogon by analogy to the mammalian hormone GLUCOGON. I think such words give insect endocrinologists a bad name, but there you have it. Trehalogon is thought to stimulate fat bodies to release trehalose into circulating hemolymph.

Releasing trehalose into hemolymph is a multi-step process that begins with hydrolysis of individual glucose residues from glycogen. This next figure shows this is under control of the enzyme phosphorylase, so named because phosphorus is involved in the hydrolysis. This enzyme yields molecules of glucose-1-phosphate, which are taken into two pathways, one is a mutase step to glucose-6-phosphate and the other is a phosphorylase step to uridine diphosphoglucose. These two glucose derivatives are combined into trehalose-6-phosphate by trehalose-6-phosphate synthetase. A phosphatase step yields the free trehalose molecule. Hence, trehalose synthesis requires hydrolysis, derivatization and assembly of two glucose residues.

Trehalose synthesis is regulated by controlling the activity of phosphorylase, the enzyme at the beginning of the pathway. There appears to be two forms of phosphorylase, phosphorylase a, the active form and phosphorylase b, the inactive form. Both of these forms exist in fat bodies, and the specific ratio of one form to the other depends upon the physiological requirements of the whole organism. Physiological requirements are expressed by circulating titres of trehalogon, the hormone active in mobilizing fat body trehalose. A final point is a reminder that the figure as shown here is taken from a recent compendium on the endocrinology of insects. It contains a slight error, shown in the phosphorylase step. This step requires the input of inorganic phosphorus, just the opposite of what is shown.

METABOLISM OF LIPIDS IN FAT BODY

We have seen that lipid reserves are stored in fat bodies as triacylglycerol. Triacylglycerol is synthesized in fat body from excess sugar and from the carbon skeletons of certain amino acids by way of the ubiquitous enzyme system fatty acid synthetase. Newly formed fatty acids are linked to glycerol backbones by acyltransferases.

Lipid is released from fat body in the form of 1,2-diacylglycerol (DAG), rather than as 1,3-diacylglycerol or a mixture of 1,2-DAG and 1,3-DAG. 1,2-DAG may be formed directly by a lipase specific to the 3-position fatty acid, or it may be formed by breakdown to 2-monoacylglycerol followed by stereospecific reacylation to 1,2-DAG. Release of 1,2-DAG is stimulated by adipokinetic hormone, but the particular enzyme target is not yet known.

URIC ACID SYNTHESIS AND OTHER DETOXIFICATION REACTIONS

The details of uric acid synthesis will be covered in the section on nitrogen metabolism and excretion. The point to emphasize here is that the fat body is the central organ of nitrogen metabolism. The storage of uric acid in urate cells and in trophocytes by some insects has been covered.

One of the great problems in crop protection by application of chemical pesticides is the natural resistance to these poisons that insects rapidly develop. Two broad mechanisms of resistance are enzymic breakdown steps, one by esterase and another by mixed-function oxidases. Under these systems, the pesticide molecules are simply metabolized into harmless forms before they can carry out their damaging effects. Most of these detoxification enzymes are formed in fat bodies and function in circulating hemolymph as well as in fat body cells.

BIOSYNTHESIS OF AMINO ACIDS IN FAT BODY

The non-essential amino acids can be formed using carbons from various intermediates in carbohydrate metabolism coupled with transaminations from preexisting amino acids. The fat body is the major site of amino acid synthesis.

AT LEAST ONE HORMONE IS SYNTHESIZED IN FAT BODY

Most homeostatic hormones, such as diuretic hormone, adipokinetic hormone and trehalogon are released from elements of corpus cardiacum or from elements of neurosecretory cells. There is an example of a hormone that is produced and released from the fat body. About 40 hours after taking a blood meal, adult female mosquitoes produce what is tentatively called host-seeking behaviour inhibitory factor. This factor is a small peptide, and is produced in fat body and released into circulating hemolymph. The exact site of action remains to be found, but one physiological effect is to switch off specific lactic acid receptors located in antennae. Fat bodies need to be physiologically prepared to synthesize the factor. Another hormone appears to be required because females that have been ovariectomized within three hours after a blood meal do not elaborate the factor.

BIOCHEMICAL ROLES OF VITAMINS

We will use this period to extend our treatment of nutrition by adding details on the biochemical roles of some vitamins. This will also serve as a review of energy metabolism, meant to help prepare for the final examination.

We recall that muscle contraction involves interactions between thick myosin filaments and thin actin filaments. These interactions occur by way of cross bridges formed by heads of the myosin molecules. Mg++ - dependent Myosin ATPase is responsible for the conformational changes in the orientation of the myosin cross bridges that result in the filaments sliding past on another. The sliding filaments result in a shortening of the Z-band to Z-band distance.

Tracheoles functionally, but not topologically, penetrate deep into rapidly metabolizing muscle fibres such that oxygen can be delivered nearly directly to muscle mitochondria, sometimes called sarcosomes. The key function of the gas exchange is to support energy metabolism during flight, or any other energy-intensive activity. By now you should have an idea of where, morphologically and biochemically, oxygen is consumed and where carbon dioxide is produced. We can revisit energy metabolism by looking back at the chart that shows types and forms of fuels.

We can also recall the metabolic pathways that yield energy to support cell work. These include glycolysis, Kreb's Cycle, the Electron Transport Chain and ß-oxidation. A physiologist might say, barring some needling details, your grasp on energy metabolism is virtually unrivalled in the western hemisphere.

The sugars, lipids and protein components that insects and other organisms consume in the course of their feeding activities are called macronutrients, not because they are particularly large or complex molecules, but because they are required in relatively large quantities. In addition to these macronutrients, insects require a host of micronutrients, so called because they are needed in more or less trace levels in the body: Some of these micronutrients are VITAMINS, which can be defined by three points. First, they are found in trace to low quantities in the body. Second, they are involved in specific metabolic functions. Third, they are not synthesized in the body. If they are synthesized, the synthetic rates do not support the organismal needs.

Some of the vitamins are directly involved in energy metabolism, acting within the pathways within our line of familiarity. We are going to look at these vitamins, and at the specific metabolic functions they perform. This is one of the areas of biochemistry that integrates common features of many different organisms. These metabolic functions are just as vital to mammals, including human beings.

Generally speaking, we recognize two major groups of vitamins - the fat soluble and the water soluble vitamins. The fat soluble group includes vitamins A, D, E, and K. The water soluble group includes the B-complex and vitamin C.

We proceed with vitamins we have already seen in action. The structure of riboflavin is shown first. This figure shows riboflavin in its vitamin form, as riboflavin, and also in its physiologically functional form in flavin adenine dinucleotide, often abbreviated as FAD. The key feature here is that the vitamin - that is, the trace substance that can not be manufactured within the body - is merely a component of a larger molecule that plays a specific role as a coenzyme. Before we discuss the role of FAD, let us introduce nicotinamide, also shown in its vitamin form and in its physiological form of nicotinamide adenine dinucleotide, similarly abbreviated as NAD.

Both FAD and NAD function as electron carriers. When an electron pair is transferred to these molecules, either one or two hydrogens are also transferred. FAD has accommodation for two hydrogens while NAD can accept one hydrogen. In NAD, then, an electron pair and one hydrogen are transferred, with a second hydrogen released into the medium. This subtle chemical difference is why the reduced form of FAD is written FADH2 while the reduced form of NAD is written NADH + H+. These compounds are particularly appropriate electron carriers because they readily accept an electron pair and they also readily donate an electron pair. We saw FAD act in the Kreb's cycle, shown in the figure. FAD accepts an electron pair from succinate, becoming FADH2. FADH2 carries the electron pair to the electron transport chain, where it releases its electrons at such position in the chain that two ATPs are subsequently formed. The figure also shows that NAD operates as an electron carrier at three places in the Kreb's cycle, carrying electrons from isocitrate, alpha-ketogluterate and from malate to the electron transport chain. Here the electrons are loaded onto the chain such that three ATPs are formed.

These examples illustrate a general feature of co-enzymes: they all provide some chemistry to enzymes that the proteins can not perform on their own. As coenzymes, FAD and NAD provide the ability to accept and donate electrons. Riboflavin and nicotinamide are necessary components of these co-enzymes, and these components can not be made in the body of insects. For perspective, recall from your biochemistry that NAD acts as a co-enzyme in more than 250 enzyme functions of this sort: carrying electrons from one place to another.

Our next water soluble vitamin is B-1 or thiamin, shown in its vitamin form and in its physiologically active form as thiamin pyrophosphate, or what is sometimes called cocarboxylase. This vitamin is quite often referred to as TPP. As a point of history, it was because of the amine group in this compound that accessory food factors were first called vitamines, an abbreviation of vital amines. Specifically, TPP is a coenzyme in the removal of carbon dioxide from various compounds. An important reaction is this one:

Pyruvate $\rightarrow$ Acetyl CoA + CO_2

Recall that this reaction is carried out by pyruvate decarboxylase. This reaction removes the carbon dioxide moiety from pyruvate, and it connects the glycolytic pathway to the Kreb's cycle.

Returning to the Kreb's cycle, note that carbon dioxide is released at two points in the cycle. The active form of thiamin, TPP, also functions in the release of this second carbon dioxide. We collect the point that some decarboxylations, or carbon dioxide removal reactions, do not depend the activity of this co-enzyme.

We have now seen three vitamins - two function as electron carriers, essentially connecting the Kreb's cycle to the electron transport chain and the third functions in decarboxylation reactions. In one decarboxylation reaction, specifically pyruvate decarboxylase, TPP also helps connect the glycolytic pathway to the Kreb's cycle.

The transport of 2-carbon acetyl units, done by Co-enzyme A or CoA, is still another key link between glycolysis and the Kreb's cycle. Recall that oxidation of fatty acids in the ß-oxidation pathway is also connected to the Kreb's cycle by transfer of acetyl CoA. This serves to underscore the importance of vitamin B-3, or pantothenic acid. This is shown in its vitamin form and as a component of Co-enzyme A. These four vitamins - nicotinamide, riboflavin, thiamin and pantothenic acid - are directly involved in energy metabolism; they act as electron carriers, in decarboxylations and as an essential component of CoA.

Insects require other water soluble vitamins in their diets, including biotin and folic acid. These do not function specifically in energy metabolism. Nonetheless, we want to briefly consider their roles in cellular metabolism. The chemical structure of biotin was worded out in 1942, and first synthesized in 1943. Biotin had been independently discovered by at least two groups, and it went under a variety of names, including vitamin H and protective factor X. Biotin functions as a carbon dioxide carrier in carbon dioxide fixation reactions. Here is an example:

$$\text{Acetyl CoA} + CO_2 \rightarrow \text{Malonyl CoA} \rightarrow \text{Fatty Acids}$$

Folic acid is actually a complex of several biologically active compounds. The formal term is pteroylglutamic acid, and it is composed of pteroic acid plus one to seven gluatmate residues. Folic acid is required for transferring single carbon units. It is essential in metabolism of some amino acids, and is probably most well known for its critical role in synthesis of thymine. Without folic acid, DNA synthesis is slowed or stopped. In the absence of DNA synthesis, production of some cells, such as red blood cells, is also slowed or stopped. This is particularly important for cells that undergo rapid turnover, such as red blood cells or insect hemocytes.

I think it is appropriate to provide a statement as to what level of detail is subject to examination. You should know the common names of the vitamins involved in energy metabolism and what their physiologically active forms are. You should be able to describe the general sort of chemistry they provide and be able to give a specific example. I would not want you to give much thought to the chemical structures of things or try to memorize many more than about half of the 250 know NAD reactions.

With respect to the water soluble vitamins, insects appear to be closely aligned with known vertebrate nutritional requirements: both groups require about the same vitamins for the same biochemical reactions. This is reasonable because the major energy yielding pathways are similar in both groups.

The fat soluble vitamins, vitamins A, D, E, and K, are not involved in the oxidative pathways we have discussed. Vitamins D and K are involved in functions occurring in vertebrates, but not invertebrates. Vitamin D is involved in bone metabolism, and vitamin K is involved in the vertebrate blood clotting cascade. Vitamin K was discovered in 1929 by a Danish scientist, Henrik Dam, who later won a Nobel prize for the work. The K comes from the Danish word koagulation. The long-standing wisdom holds that insects or other invertebrates do not require vitamins D and K.

Vitamin A is involved in many mammalian tissues, mostly in ways not completely understood. One well-appreciated action of this nutrient is its role in vision. We will discuss the physiology of vision in a later lecture, in which the role of vitamin A will be addressed in detail. For now we will just make the point that vitamin A is required for insect vision. Vitamin E is also thought to be dietarily essential in at least some insects. Vitamin E is essential for fertility in rats, and apparently in male crickets. The specific roles of vitamin E are not understood, although its most well known activity is its action as an antioxidant. Vitamin E is thought to interact with certain free radicals, and thereby prevent oxidative damage to cells. Whether vitamin E is required by all insects is difficult to say. I speculate it is required, however, the requirement is very hard to detect because the vitamin is stored in tissues and turns over only very slowly.

Now there is one last problem in energy metabolism. Referring to the glycolytic pathway again, please note that at one point NAD picks up a pair of electrons to become $NADH^+$ H^+. There are two points to be made: One, if the electrons were not subsequently donated from the NADH, cells would eventually become depleted of NAD - and energy metabolism would come to a screeching halt right at this step. So it is important to recycle this electron carrier to keep energy metabolism rolling along. The second point is that this pair of electrons represents a certain amount of potential energy in the form of additional ATPs if the electron pair could be shuffled over to the electron transport chain. Stated in problematic terms, we have this problem: the outer mitochondrial membrane is not permeable to NAD. There are two pools of NAD in most cells. One is the inner mitochondrial pool and the other is the cytoplasmic pool. Since the electron transport chain is tucked away in the inner membrane of the mitochondria, how is the electron pair, associated with the cytoplasmic pool of NAD, to be delivered to the electron transport chain, or is it delivered at all?

Let us consider the operation of the glycerol-3-phosphate shuttle. On the left is the glycolytic pathway. This is in the cytosol fraction of the cell, where the NAD accepts an electron pair and becomes NADH (plus H^+). The electron pair is transferred to dihydroxyacetone phosphate (DHAP). This step regenerates the NAD, thereby allowing the glycolytic pathway to continue its important operations. Reducing the DHAP by addition of electrons converts it from DHAP to glycerol-3-phosphate. This newly formed compound easily migrates across the outer mitochondrial membrane. Again, the membrane is a barrier to NAD flux. Once in the mitochondria, the electron pair is donated to the electron transport chain at the level of FAD, which converts the glycerol-3-phosphate back into DHAP. DHAP is free to wander back out into the cytosol. Then two points are covered by the glycerol-3-phosphate shuttle: the NAD is regenerated and the potential ATPs are realised.

This shuttle is best known from insect flight muscles. There are other pathways to account for moving reducing equivalents from cytosol into mitochondria, one of which is called the MALATE SHUTTLE because malate is formed by oxidation of oxaloacetic acid. The malate moves into the mitochondria, where it is converted to fumarte and then into succinate. This is a more complex shuttle, and I once speculated at a scientific meeting that the glycerol-3-phosphate shuttle is much faster than the malate shuttle, hence more appropriate to very rapidly metabolizing tissues, such as insect flight muscle.

Chapter 3

Transport of Energy

The major pathways of oxidative metabolism, namely glycolysis, ß-oxidation, Kreb's cycle and the electron transport chain. We should be familiar with where these energy-yielding pathways are located within cells, and what principle substrates are taken into these pathways. Let us expand our information just a bit, by introducing the idea of metabolic scope. Rates of metabolism can be measured by oxygen uptake rates. When oxygen uptake is measured at rest and during rigorous activity the range between these minimal and maximal metabolic rates is called the metabolic scope. The point to be made is that insect flight muscle tissue can be characterized as having the largest metabolic scope in the animal kingdom. Here are some examples:

Metabolic rates during flight are about 4 to 6 times higher than metabolic rates at rest in hummingbirds. The actual values are about 10 to 14 ml oxygen /gram body weight/ hour at rest and 40 to 85 ml oxygen /gram body weight/ hour at rest. By comparison, metabolic rates during flight in some insects can be 200 times the resting rates. Some butterflies increase their metabolic rates from 0.5 to 100 ml oxygen /gram body weight/ hour when they begin flying. These data just make the point that differences between resting and flight metabolic rates - that is, metabolic scopes - are much larger in insects than they are in even the most rapidly metabolizing vertebrates - hovering birds. Such large metabolic scopes pose the problem of supplying fuel to the flight muscles such that the hemolymph supply responds to changes in metabolic demand. We can imagine some odd things.

1. Soon after initiation of flight the hemolymph reserves of fuels are used up. If more fuel is not mobilized from the fat body and released into the hemolymph in a timely way, the insect may effectively "run out of gasoline". Who knows what would happen after that.
2. At another extreme, after flight activity has ended, if fuels were continuously released into the hemolymph at rates sufficient to support the necessarily high rate of energy metabolism involved in flight, the circulatory system would soon become so clogged up as to effectively block hemolymph movement.

The release and transport of metabolic fuels must, perforce be regulated in relationship with metabolic demands. The topic of this discussion is the hormonal, or endocrine, control of hemolymph fuel levels.

The circulatory system of insects is an open system. Hemolymph is aspirated into the heart, or dorsal vessel in the abdominal segments and pumped forward, through the thorax and on into the head. Hemolymph flows out of this single circulatory tube into the open hemocoel and percolates back through the body tissues to the abdomen for recirculation. Hemolymph circulation is driven by contractions of muscles connected to the dorsal vessel and by volume changes in the abdomen and thorax. This circulation is responsible for the transport of fuels and for transporting various hormones and other nutrients throughout the organism.

One of the most thoroughly studied cases of endocrine regulation of fuel transport in the hemolymph is the role of adipokinetic hormone (AKH) in the desert locust. As the name suggests, this is a lipid mobilizing hormone.

Early on in the study of flight energetics it was shown that after locusts fly for more than a few minutes there is an increase in the lipid content of the hemolymph. The increased lipid is mainly in the form of diacylglycerol (DG) as a transport form of metabolic lipid. A key point is this increase in lipid concentration occurs even though there are considerable increases in the rate of lipid metabolism in flight muscles and in the rate of lipid uptake from circulating hemolymph. We understand this to mean that even though lipids are being rapidly removed from the hemolymph by transport into flight muscles there is a net increase in hemolymph lipid concentration. It appeared from these observations that the hemolymph concentration of lipid is regulated. The regulatory mechanism was guessed to be hormonal regulation.

The first observations showed that hemolymph lipid concentrations increase after some minutes of flight in the locust. The first endocrinology experiments were to surgically remove several tissues from locusts, then create aqueous extracts of the tissues. These extracts were individually injected into non-flying locusts. Extracts of most tissues did not induce an increase in hemolymph lipid levels. When extracts of an endocrine organ, the corpus cardiacum (CC), were injected into resting locusts, an increase in hemolymph lipid concentration followed. The increases occurred in a time course similar to the elevation in lipid that was regularly seen as a result of sustained flight activity. Furthermore, when the CC extracts were applied to isolated fat body tissue *in vitro*, DG was released into the incubation medium. These observations provide the basis of a classical endocrine system: an endocrine gland releases a factor into the circulatory system. The factor is carried by way of the circulation to some target organ distant from the endocrine gland.

People moved from these organismal observations to trying to characterize the endocrine factor. The active principle was found to be stable to boiling for an hour or so; stability to boiling is a property of small molecules. The activity of CC extracts was destroyed by treatment with proteolytic enzymes; this is interpreted to suggest the factor is a peptide or protein. Combined with the heat stability, it was thought the factor was a small peptide. It was also found that hemolymph from resting locusts contained no traces of the factor, while hemolymph from locusts that had been flying for the last 90 minutes or so had small, but detectable quantities of the factor. The factor began to look very much like a real hormone.

In 1976 an English group succeeded in purification of the hormone, then went on to determine the chemical structure. It turned out to be a peptide hormone composed of 10 amino acids. This was the first complete identification of an insect peptide hormone, then named adipokinetic hormone (AKH).

The only one done so far, but that is no longer even close to true. Several groups worldwide now work on peptide hormones in insects. Usually people try to isolate the factor, then determine amino acid composition, then get amino acid sequences. Two things have helped speed this business along. First is development of very beautiful columns that can be used to separate peptides on HPLC. The second thing is development of peptide sequenators than can work with as little as a few picograms of material. Between the HPLC for isolation of peptides and the sequence technology, the endocrinology of insect peptide hormones is developing into a wonderfully productive area of insect biochemistry and physiology.

Back to AKH. The next step was to confirm the structure of the peptide. The generally accepted approach to confirmation is to synthesize the peptide, and then show that the biologically activity of the synthetic product is identical to the activity of the naturally occurring hormone. At this juncture, we saw a biological phenomenon, namely the increase in hemolymph lipid titres after some time in flight activity, and some of the experiements supporting the idea that hemolymph lipid concentration is under endocrine control.

Before going on, we need to review the anatomy of the main endocrine organs in insects. The important landmarks include the brain with optic lobe, the pars intercerebralis (PI) and two sets of neurosecretory cells - median neurosecretory cells and lateral neurosecretory cells. Neurosecretory cells resemble typical nerve cells or neurons, with elongate axons, but they are characterized by microscopic evidence of secretion. The secretion is granular with characteristic staining properties in histological preparations.

The visible granules are probably only large carrier molecules with a much smaller hormone molecule attached. Histological studies have allowed construction of the following series of events: peptide hormones are produced in the cell body, then attached to a large carrier protein and moved down the axon process for export at the terminus of the neurosectretory axon. When the hormone is released it is separated from the carrier molecule and enters the circulating hemolymph. This is a common mechanism to release hormones from the brain.

Let's consider a cross-section of the CC from the locust. The CC are paired endocrine glands that often form part of the aorta. The CC contain nerve endings from neurosecretory cells of the brain. Part of the CC store and release hormones from these neurosecretory cells from the brain. In this storage and release function, the CC are properly called neurohemal organs. Association with the aorta is thought to facilitate broadcast of hormones because the hormones can be directly secreted into circulation.

The CC also contain intrinsic secretory cells. Unlike the neurohemal portion of the CC, these cells produce and release their own hormones. Sometimes the neurohemal and intrinsic parts of the CC are intermingled, but in the locust they occur as separate lobes. Because these lobes could be surgically separated, it was possible to show that most, and probably all, of the AKH of the locust CC was present in the intrinsic rather than neurohemal lobes. We conclude that AKH is produced in the CC rather than in the brain. This idea was tested by removing the intrinsic lobes from locusts, then allowing them to fly. The surgery prevented the normal rise in hemolymph lipid that accompanies flight. Also, cutting the nervous connections between the brain and the CC similarly blocked the normal rise in hemolymph lipids. This suggests that nervous stimulation from the brain triggers the release of AKH by the CC.

We might imagine a control system as follows: prior to a lengthy flight a locust has a standing hemolymph reserve of trehalose and DG associated with lipophoren. During the first few minutes of flight, the carbohydrate is transported into the flight muscles and oxidized. This is followed by a shift to lipid oxidation. Sometime during this shift the brain sends a nerve impulse to the CC, resulting in the release of AKH which finally circulates to target cells in the fat body. The fat body cells are triggered to mobilize storage triacylglycerol and release DG to lipophorin in the hemolymph.

This is a reasonable enough system, but there is evidence that AKH may also regulate the balance of carbohydrate and lipid oxidation in flight muscles. The first line of evidence is that the release of the hormone from the CC appears to be at least partly regulated by the trehalose concentration in the hemolymph. Injection of trehalose into locusts to create above-normal sugar levels inhibits AKH release and lipid mobilization in flown locusts. The second point is that AKH seems to act on the flight muscles directly to influence oxidation of lipid over carbohydrate.

So, we modify our picture to think that a drop in hemolymph sugar is sensed in the brain, followed by a release of AKH from the CC which acts on the fat body to mobilize and release lipid and acts on the flight muscle to stimulate a shift in substrate utilization.

Now, just as we think we are beginning to get a grip on things, we introduce a third role for AKH. In resting locusts the hemolymph lipids are associated with lipophorins, known as high density lipophorin (HDLp). HDLp is also involved in transport of lipids from the site of absorption in the alimentary canal to the fat body and in the transport of sterols and lipoidal hormones. Aside from lipophorins, there is another hemolymph protein first called C2 by one group and finally named apolipophorin-III (apo-Lp-III) by another group. It is called apo-Lp-III because the major lipophorin, HDLp, is composed of two subunits called lipophoren-I and lipophorin-II. Without further concern for nomenclature, here is how the system works.

Following injection of CC extracts, or of synthetic hormone, or after a period of flight, additional DG is found in the hemolymph associated with a new lipophorin, low density lipophorin (LDLp); at the same time the amount of apo-Lp-III in the hemolymph is depleted. It is thought that during lipid mobilization some apo-Lp-III combines with HDLp to produce LDLp. This form has greater lipid carrying capacity than HDLp. Here is a bit of data on the carrying capacity of the various forms of Lp.

Table: Properties of high density and low density lipophorins

Lipophorin	Lp Density (g/ml)	Ratio of Lp1: LpII:apoLpIII	DAG content (g/mmol Lp)
HDLp	1.11	1:1:1	62
LDLp	1.04	1:1:14	390

Summarizing, we imagine that AKH has three physiological roles: mobilization of fat body lipid reserves, a shift in flight muscle oxidative balance to favour use of lipids and a change in lipophorin pattern to allow increased lipid carrying capacity.

While most of what we know about AKH comes from work on locusts, there is also evidence for similar hormones in other insect species. The tobacco hornworm is another model system for these studies and also the beetle *Tenebrio* was reported to have an AKH response to CC extracts. The migratory monarch butterfly, which relies on lipid reserves for its long flights, shows elevated hemolymph DG upon injection of CC extracts.

On the other hand, as we have said before, insect physiologists are of necessity comparative physiologists. This is because insects as a group have evolved many different physiological and biochemical adaptations to similar situations. As comparative physiologists we must take care not to generalize too broadly. For example, injection of CC extracts from either the cockroach *Periplaneta americana* or from *Locusta* into cockroaches causes a decrease in hemolymph DG. Conversely, injection of cockroach extracts into locusts results in an AKH effect that matches the locust AKH effect. At this point I would speculate that most insects have an AKH, but details of the broader physiology remain rather fuzzy.

One of the criticisms that frequently arises in hormone research is the assessment of the physiological relevance of hormones. For example, a high dose of AKH from locusts can cause a hyperglycemic effect in cockroaches. A hyperglycemic effect is an increase in hemolymph sugar concentrations. These sorts of non-specific effects of high hormone dosages are called pharmacological as opposed to physiological effects. The question, then, is what information do we need to critically assess physiological activity of a hormone?

The question is quite important because many people are able to isolate peptide factors in a fairly straightforward way. The problem is, the isolators have no idea relative to the biological meaning of their factors. The isolators usually fnd some collaborators to help, and newly isolated factors are routinely tested by bioassay. It is best if you can send your factors to several people. Somebody might test the adipokinetic hormone activity of your factor; somebody else might test the diuretic hormone activity in a completely different system. My colleague, David Petzel, once tested the diuretic hormone activity of a cockroach factor - using mosquito Malpighian tubules. He got a positive response alright, but was his response a physiological activity of a hormone?

Let us consider these points:

Points in Evaluating the Physiological Activity of a Hormone

- The chemical identity of the proposed hormone
- Presence of the hormone in an endocrine gland

- Specific response by a specific target tissue or tissues the hormone is thought to effect
- Demonstration of a dose-dependent response
- Release of the hormone into the hemolymph in response to a specific physiological stimulus
- The presence of the hormone in the circulatory system in some appropriate physiological circumstance and concentration

Technical problems can be counted upon to make such complete information impossible to obtain except in rare cases. Let us turn again to AKH, where research on the locust system is most complete. We attempt a critical assessment. AKH has been purified and its chemical structure determined:

Glu-Leu-Asn-Phe-Thr-Pro-Asn-Trp-Gly-Thr-NH_2

The structure has been verified by peptide synthesis and recording the biological activity of the synthetic peptide. Surgical experiments showed the hormone is present in the intrinsic portion of the CC. Also, a specific target tissue, the fat body, responds to the hormone in a dose-dependent manner. The hormone has been found in the circulating hemolymph under appropriate physiological conditions - that is, it appears in the hemolymph of locusts after about 90 minutes of flight, but not in resting locusts. The hormone appears to be released in response to a nerve impulse from the brain as a result of changing hemolymph trehalose concentrations. This last point is still a little bit weak; nevertheless, the available information on AKH permits judgement that this is, indeed, a physiological endocrine system.

RESPIRATION INAQUATIC INSECTS

One of the major issues in the physiological ecology of aquatic aerobic animals is the availability of oxygen. Compared to atmospheric chemists and other people along those lines, physiologists take a rather simple view of air. We generally consider air is composed of something like this:

Element	Proportion
Oxygen	21 per cent
Nitrogen	79 per cent
Carbon dioxide	0.03 per cent

We can quickly gather a couple of points from this information. First, there are several trace elements in air that are lumped into nitrogen for purposes of discussion. This is acceptable because the quantitatively minor gasses, such as argon, are also inert, and physiologically not different from nitrogen. The second point is that we can note from the percentage composition that a liter of dry air at sea level contains about 210 ml of oxygen. Just from watching animals moving around on land, 210 ml of oxygen per liter of air seems to be plenty of oxygen.

Well, Ok, let us ask how much oxygen is in water? The answer is not quite so simple, even for physiologists. To correctly look at availability of oxygen to animals living in water, we need to consider several points. First, gases are not equally soluble in water. Hence, the solubility of individual gases will partly determine availability of oxygen. Here are a few data on solubility of pure gases in distilled water at 15°C and 760 mm Hg (1 atmosphere of pressure):

Element	Solubility
Oxygen	34.1 ml/liter
Nitrogen	16.9 ml/liter
Carbon dioxide	1019.0 ml/liter

We can use these solubility data to calculate volumes of gas available in water when the water is in equilibrium with air at one atmosphere:

Volume of gas = Solubility X % Composition

Element	Solubility x % composition	Volume
Oxygen	34.1 × 21%	7.16 ml/liter
Nitrogen	16.9 × 79%	13.35 ml/liter
Carbon dioxide	1019 × 0.03%	0.31 ml/liter

Now with just a little bit of high-powered mathematics, such as long division, we can calculate that the 7.16 ml/l of oxygen in water is about 30 times less than the 210 ml/l of oxygen in air. We can also calculate the percentage composition of gases in water at 15°C and one atmosphere pressure by adding the three volume values and dividing each by the total:

Element	Volume/total	Percentage composition in water
Oxygen	7.16/20.82	34%
Nitrogen	13.35/20.82	64%
Carbon dioxide	0.31/20.82	0.01%

These values hold for water at 15°C, equilibrated with air at one atmosphere, typically regarded as sea level. The amont of gas in water, however, varies with water temperature, among other physical parameters. The following table will make the point:

Water temperature (°C)	Volume of oxygen (ml/liter)
0	10.3
10	8.0
15	7.2
20	6.6
30	5.6

This table shows an inverse relationship between water temperature and oxygen contents, that is to say, there is a constraint on oxygen availablity in water due to the influence of temperature. This temperature constraint presents a problem because most aquatic animals, particularly small animals such as insects, are poikilothermic. The body temperatures of poikilothermic animals vary with ambient temperature: as water temperatures increase, so do body temperatures.

The problem stems from the influence of temperature on metabolic rates. Metabolic rates are directly related to body temperatures. As metabolic rates increase, oxygen consumption rates increase. The environmental influence of water temperature is seen in body temperature, which influences metabolic rates. These factors intersect at a crucial, yet paradoxical point: increasing oxygen requirements are attended by decreasing availability of oxygen in water.

Respiration in aquatic animals, then, is a look at physiological solutions to the problem of obtaining sufficient oxygen to support metabolic needs from an environment where oxygen is in short supply. From the perspective of environmental scientists, we note oxygen is one of the physical parameters of surface fresh water that change markedly in reaction to the various of the polluting chemicals regularly dumped into lakes, rivers and streams. Many aquatic insects and certain of our favourite fresh water fishes can not survive in waters with reduced oxygen levels. The presence or absence of these animals that require higher levels of disolved oxygen can be used to assess the relative environmental health of a water system. Such species are referred to as indicator organisms.

More germane to our topic, however, is understanding that insects and other aquatic animals have potential problems of low oxygen availability. One solution to the problem of obtaining oxygen while living in aquatic environments is to directly breathe atmospheric air. Many insects have respiratory siphons. In insects with this sort of anatomical arrangement only the posterior spiracles are functional, and they are located at the end of a siphon such that only the posterior portion of the insect penetrates the surface film of water, while the rest of the body is submerged, often suspended from the surface film. The respiratory siphon on larvae of drone flies, *Eristalis*, is telescopic, not meant to suggest one can see through it. Mosquito larvae also have respiratory siphons, but they are not so long as in drone flies.

One constraint on the function of respiratory siphons is that the tips of siphon tubes must be arranged such that air can move into the siphon while water is excluded. Commonly, siphons have hydrofuge hairs associated with their spiracles.

When their spiracle is submerged the hairs close over the spiracle and water is kept out of the siphon. At the surface, the hairs are separated by surface tension forces, allowing air to pass to and fro within the siphons. There is typically a funnel shape at the surface.

Respiratory siphons have certain shortcomings. The siphons tend to limit insects to the upper levels, or strata, of water, so that they are close to the surface. Insects require other adaptations to achieve some independence from the surface strata. Some insects manage to exploit deeper reaches of water by carrying air stores along with them during their dives. This strategy is well developed in the predacious diving beetle *Dytiscus*. By keeping air stores in contact with spiracles during a dive, these beetles have their trachael systems plus air stores to draw upon to meet oxygen requirements.

Spiracles are typically located along the sides of many terrestrial insects, but this arrangement does not help much with carrying supplies of air under water. Many beetles have their abdominal spiracles on the dorsal surface, under the elytra. This needs be only slightly adjusted for carrying air stores during diving. Many other insects, especially aquatic true bugs also carry air stores under water.

The efficacy of the air store is vastly improved because the bubble itself can act as a physical gill. That is, it can extract oxygen directly from the surrounding water. The following work on the backswimer *Notonecta* illustrates the point. When this bug was held under water that had been equilibrated with pure nitrogen, it survived for only 5 minutes. When the water was equilibrated with air the insect could survive under water for 6 hours, 72 times as long. These two observations are consistent, and they can be understood with respect to extraction of oxygen from the water. The importance of nitrogen in the air bubble can be shown by the following experiment. If *Notonecta* is given a gas bubble of pure oxygen, instead of an air bubble, the bug can survive for about 35 minutes under water. The role of nitrogen in performance of a physical gill can be explained as follows:

Soon after a bug dives, a disequilibrium occurs. Removing oxygen from the bubble by metabolism decreases the partial pressure of oxygen compared to water and increases the partial pressure of nitrogen. Oxygen tends to move down its concentration gradient into the bubble while nitrogen moves down its gradient from the bubble into water. The solubilities of gases become important. Nitrogen has low solubility in water, and it diffuses out of the bubble more slowly than oxygen diffuses from the water into the bubble. The nitrogen is essential to the bubble as physical gill because without it there would be no change in partial pressure of oxygen; the differential partial pressures drive oxygen from the water into the bubble.

This system can provide enough oxygen to small, inactive insects to allow them to remain submerged for months. More active insects require more oxygen. The diving beetles have to surface every few minutes to replenish their oxygen stores. In addition, certain physical parameters assert themselves. For one, as we saw earlier, water temperature effects the amount of oxygen that can be dissolved in water. At higher temperatures there is less oxygen in water, and at the same time an insect requires, and consumes, more oxygen. A physical gill may be of small value under such circumstances. Another factor impacting on air bubble life spans is depth in water. For every meter of depth in water there is about a tenth of an atmosphere increase in pressure. Increased pressure speeds diffusion out of air bubbles, and shortens their useful life span. Here is a summary of several factors that impact on the useful life spans of air bubbles, and thereby determine how long an insect may survive under water:

1. Temperature of water
2. Amount of O_2 in water
3. Metabolic rate
4. Size of bubble
5. Depth of dive

Now, when an insect dives with an air bubble, nitrogen slowly diffuses into the water, and the bubble becomes progressively smaller until it collapses. The collapse is delayed because nitrogen moves into water only slowly, but sooner or later, air bubbles will collapse. Insect using air bubbles as physical gills must return to the surface. If an air bubble were non-collapsible, many aquatic insects would not have to surface. Some insects, such as the water bug *Apohelocheirus* are equipped with a plastron, which holds a non-collapsible thin film of air along the outer surface of the body. A plastron is continuous with the trachael system. A plastron is a dense (up to 2.5 million hairs/ square mm of surface area) layer of stiff hydro-phobic hairs.

Plastrons typically run along the body surface, continuous with the spiracles. There are cuticular openings which lead into air channels, which in turn lead to spiracles and trachael systems. The hydrophobic quality of hairs along the air channels serve to keep water out of trachael systems.

The most efficient arrangement of plastron hairs would be to have them parallel to a body surface. Insects approximate this by having the tips of plastron hairs bent, or else they slope as some angle, so that air films resist collapse when subjected to higher pressures. Plastrons function in the same patterns as air bubbles when they are extracting oxygen from water, that is to say, they act as physical gills. Plastrons are found on surfaces of many insect eggs and on many species with aquatic pupal stages.

We have seen two approaches to organismal respiration in aquatic environments. One is to breathe air directly by way of siphons and another is to extract oxygen from water using a sort of physical gill. Several groups of insects express a third approach, which is to develop a true physiological gill to extract oxygen from water. The term gill, with no prefix such as "true" or "physiological" is sufficient because other sorts of gills are linked to an appropriate modifier, such as anal gill or physical gill. Gills are appendages for gas exchange, and tracheae branch into these appendages. Oxygen diffuses from water into tracheae of gills, and from there moves under the usual motivation by way of trachael systems to tissues.

The locations of gills varies in larvae of several insect species. Stone fly nymphs, *Pteronarcys*, have gills on the thorax and first two abdominal segments. Gill plates are located at the end of abdomens of damselfly larvae. Finally, mayfly nymph have gills on thoraces and abdomens. Just in passing, the gills of mayfly nymphs are equipped with nervous and muscular connections. The gills can be swept back and fro to mix water, and thereby alleviate the problem of local oxygen depletion. This arrangement is the source of the tracheal gill theory in arguments over the evolution of insect flight.

Gills extract oxygen from surrounding water. Due to its low diffusion rate in water, oxygen can be depleted from water immediately local to respiratory surfaces. This local oxygen-poor water must be replaced by fresh, oxygen-rich water. Moving water for this purpose is called ventilation of the gills. Some insect ventilate their gills by selecting a microhabitat where water is flowing. Still others ventilate by muscular action. This can require considerable energy expenditure because water has greater density and viscosity than does air. The oxygen costs of ventilation must be more than offset by oxygen gains. Work on caddis fly larvae illustrates the point. In the experiments, metabolic rates of caddis fly larvae increased when water current velocities decreased. These data suggest the larvae can somehow register changes in oxygen availability, then respond by ventilation. You can try it yourself. Hold your head underwater for some time, see if you don't register a change in oxygen availability.

Insects use a variety of methods to ventilate their gills. Mayfly nymphs beat their abdominal gills in the absence of water movement. Stonefly nymphs ventilate by doing sort of "push-ups" at a rate related to amounts of oxygen in water. Larval dragonflies combine ventilation of gill surfaces with locomotion. Dragonfly larvae aspirate water into their rectums, which feature rectal gills. These larvae can locomote by jetting themselves forward. They do this by rapidly squirting aspirated water back out of their rectum. Damselfly larvae use their abdominal gills as paddles to propel themselves while simultaneiously ventillating their gills.

In vertebrate respiratory systems the respiratory pigment hemoglobin serves three functions. First, it withdraws oxygen from the air at respiratory surfaces in lungs. Second, it transport oxygen throughout bodies. Third, hemoglobin releases oxygen to tissues at controlled rates. Some insects also contain hemoglobin molecules, but the role of insect hemoglobin is storage, rather than transportation, of oxygen. Insect hemoglobin is about half the molecular weight of vertebrate hemoglobin, and it has a much higher affinity for oxygen. That is, it can be saturated at much lower oxygen tensions. Finally, insect hemoglobin is disolved in hemolymph; it is not carried in blood cells.

The larvae of certain midges living in anoxic mud have hemoglobin in their hemolymph. To emphasize the point, these larvae tend to look sort of pink when they are well oxygenated. They live in burrows, in which they periodically wiggle, setting up currents of water. The fresh water contains oxygen, which is taken up by the insect hemoglobin. The hemoglobin can store enough oxygen to support about 9 minutes of metabolism.

Two aquatic hemipterans, *Buenoa* and *Anisops,* also store use hemoglobin to store oxygen. Part of their oxygen stores for dives is in a small bubble, the rest is in hemoglobin. Their bubbles are small enough the insect can have neutral buoyancy and remain suspended in water while awaiting prey. It has been suggested that with larger bubbles they would not be able to suspend themselves in water, and without hemoglobin they would have to surface much more often. In this case, the hemoglobin seems to be an adaptation for predation in water.

Finally, another adaptation for obtaining oxygen while remaining submerged is to extract oxygen from aquatic plants. Some insects do this by thrusting their spiracles into the aerenchyma of plants; their spiracles are located at the tip of a sharp-pointed post-abdominal siphon. Larvae of two dipterans, *Chrysogaster* and *Notiphila* do this, as well as larvae of the beetle *Donacia*. These insects live in oxygen-poor mud.

Chapter 4

Insects Immunology

HUMOURAL IMMUNITY

One question entomologists and university administrators might wonder about now and again would be along the lines of "What are some of the emerging areas of insect physiology that are likely to have great impact on how entomology might be carried out in the future? Several topics come to mind, including understanding the roles of peptide hormones, the significance of biogenic amines, more detailed appreciation of sensory physiology, and intracellular signal transduction mechanisms.

However, above all of these topics, and this is certainly not to be taken as an exhaustive list, the study of insect immunology may well develop into one of the most active and most important areas of insect physiology.

INSECT DISEASES

The idea that insects can have diseases goes well back to antiquity. The idea predates the science of microbiology by centuries. The earliest records come from descriptions of a pest of honey bees by Aristotle. In the infancy of microbiology, Pasteur showed that two diseases of silkworms were caused by microbes. As you might expect, much of Pasteur's work was motivated by economic concern: the silk industry suffered considerable losses caused by diseased silkworms.

In more recent times insect pathogens have used microbes as agents of biological control of insects. Although many microorganisms have been studied as actually or potentially useful control agents, none have so far received as much attention as *Bacillus thuringeinsis*. The potential of various microbial forms, including bacteria, fungi and protistans, as insect control agents has driven an impressive body of work on insect microbiology and pathology.

The study of insect immunity went along with much of the very early work on the general question immunology. There was a period of considerable activity in the 1920's. A second surge of work on insect immunity started in the late 1950's, when it was discovered that a certain immunity could be induced in some insects. This induced immunity can be compared to a sort of standing, or constitutive, immunity. The most recent period of research on insect immunity began in 1980 with the first isolation and purification of an induced antibacterial factor. Modern work on insect immunity is particularly satisfactory because it is one of the areas of insect biology where molecular biological tools have yielded a rich harvest of information that would not otherwise have been generated.

It is appropriate to define immunity. Immunity is "resistance to or protection against a specified disease". This definition draws attention to the phenomonology, and it is independent of specific immunological mechanisms. The distinction is important because the contemporary wisdom in some circles holds that insects do not have an immune system because they do not produce antibodies of the mammalian sort.

THE MAJOR IMMUNITY MECHANISMS

There are two major areas of insect immunity, cell-free immunity and hemocytic immunity. These are sometimes called humoural immunity and cellular immunity, respectively. These two immunological responses are complimentary, and both may be seen in response to the same infection. Please note that immunity in insects and in other invertebrates differs from the immune systems of mammals in fundamental ways. First, insects produce no lymphocytes, of which there are many sub-types including eosinophils and neutrophils, no T-cells and no helper cells. So the cellular immunity of insects is not to be compared to the cellular immunity of mammals, although some people have tried to make that sort of comparison. Second, insects do not biosynthesize immunoglobulins, such as gamma-globulin. The important point here is that the immunity of insects lacks the molecular specifity of antibodies from mammals.

INDUCED ANTIBACTERIAL PROTEINS

Wounding or injection of certain bacteria some insects induces the biosynthesis of antibacterial proteins called cecropins. These proteins were first isolated from the hemolymph of *Hyalophora cecropia,* from whence the term cecropin was coined. Molecules with similar structure have since been isolated from the waxmoth *Galleria,* the tobacco horn worm *Manduca sexta,* several species of Diptera and from a beetle. Not all insects appear to produce cecropins, because a cricket, a cockroach and a locuts do not produce them. The blood-sucking bug *Rhodnius* produces antibacterial proteins that differ from cecropins. *H. cecropia* produces cecropins A, B and D as well as several minor cecropins, C, E, and F, which differ from the major cecropin by only a single amino acid substitution. All of these proteins are about 4000 in molecular weight, each with 35 to 37 amino acids. There is a very high level of homology among these proteins, all of which feature long hydrophobic reaches in the C-terminal portions of the molecule. The very close homology among these proteins is consistent with the idea that they have arisen through gene duplication.

The cecropins are not active against many, but not all, bacteria. Cecropins A and B are active against a large number of gram-positive and gram-negative species, including some that are occasional insect pathogens and some that are obligate insect pathogens and others that are non-pathogenic species. None of the cecropins are active against *B. theringeinsis.*

ANTIBACTERIAL ACTIONS

Cecropins lyse bacterial cell membranes; they also inhibit proline uptake and cause leaky membranes. These actions at the level of procaryotic membranes are consistent with the structures of cecropins. They are all cylindrical, amphipathic molecules with long hydrophobic regions on one end. In effect, these protein act like detergents. Synthesis of proteins that are analoges to cecropins show that activity against a broad spectrum of bacteria species is lost when amino acid substitutions that disrupt the alpha-helix structure are made.

The genes that code for cecropins have been cloned. These molecular biological approaches give insight into the biosynthesis of the cecropins. The gene for cecropin B, for example, codes for 26 amino acids in the N-terminus of the protein that are removed during post-transcriptional processing to produce the mature cecropin B. These are removed in 3 steps, first 22 residues are removed in one cleavage, which produces a pro-cecropin. The pro-cecropin is further processed by two sequential cleavages that yield the final product.

ATTACINS

A second set of antibacterial proteins are called attacins. These also have been isolated from *H. cecropia* and three other moth species. Attacins are considerably larger than cecropins, made up with over 180 amino acids. The antibacterial activity of these proteins covers a rather narrow spectrum compared to the cecropins, with high activity only against a few species of bacteria in the alimentary canal. Their site of action is at the outer bacterial membrane. These proteins appear to work in synergism with the cecropins and with lysozyme so that the three types of immune proteins can work together.

LYSOZYME

Peptidoglycan is a large, bag-shaped macromolecule that encases bacterial membranes. It is universally present in bacteria. Minor variations of peptidoglycan occur in among different bacterial species. Lysozyme is an enzyme that hydrolyzes the glycosidic bonds of peptidoglycan, and thereby causes bacterial cell lysis. This enzyme constitutively occurs in low levels in hemolymph of most insects. Lysozyme is rapidly induced immediately upon bacterial infection.

STILL OTHER ANTIBACTERIAL PROTEINS

Antibacterial proteins widely occur in vertebrate and invertebrate animals. It is not too surprizing, then, to learn that representatives of various insect orders produce different antibactieral proteins. The attacins and cecropins are known mostly from lepidopterans, although they have been detected in other insects and even in intestine of pigs.

Honeybees produce another family of such proteins that are known as apidaecins. These are small peptides, molecular weight about 2100. Apidaecins do not appear to disturb bacterial membranes, and it is not yet clear how they do work. It is speculated that these are bacteriostatic, rather than bacteriocidal, proteins. Royalisin is another antibacterial peptide is found in the royal jelly of honeybees. This is also an amphipathic protein, and its mode of action may be similar to cecropins. It is a small molecule, molecular weight estimated at 5523.

Dipterans also produce induced antibacterial proteins. Two peptides from *Phormia* are called defensin and diptericin. Another dipteran peptide is a male-specific antibacterial product from *Drosophila melanogaster*. This peptide is known as andropin, and it thought to protect seminal fluid and the male reproductive tract against microbial infections.

One of the most exciting findings in insect immunology is the discovery of an insect immune protein called hemolin. This protein belongs to the immunoglobulin superfamily, that is, it shares sequence homologies to mammalian immunoglobulins. It is thought that this protein is one of the first proteins to appear in the hemolymph of the giant silkmoth *H. cecropia*. This protein binds to the surface of bacteria, and it is likely that the binding may be the first step in the insect immune response.

BACTERIAL RESISTANCE TO INSECT IMMUNITY

With respect to evolutionary ecology, our discussion of insect immunity offers a sort of physiological perspective on host-parasite relationships. Within the context of these relationships, we can imagine strong evolutionary pressure on the microbes to develop mechanisms that circumvent immunity reactions with hosts. Two such mechanisms are known in some of the bacteria that invade insects. These are called passive resistence and active resistence. Passive resistence is related to the structures of some bacterial envelopes. These envelopes happen to be resistent to antibacterial activity of immune proteins. In these cases the resistence is a feature of outer membranes, rather than something bacteria might actively produce. There appear to be mutants of insect-pathogenic bacteria that have lost this sort of passive protection and are comcominantly less virulent.

Active resistence refers to active biochemical breakdown of immunity, rather than a sort of side-stepping as seen in passive resistence. B. theringeinsis, for example, produces an immune inhibitor called InA that works by proteolytic breakdown of cecropins and attacins. A beautifully complex relationship occurs in a nematode with a symbiotic bacterium appropriately named *Xenorhabdus nematophilus*. This bacterium is sensitive to two cecropins, A and B. However, the nematode produces an immune inhibitor enzyme that digests cecropins and attacins, thereby rendering them without activity.

One co-evolutionary response to these bacterial proteases that inactivate the insect anti-bacterial proteins may be another protein produced by some insects that inhibit the bacterial proteases. A molecule known as alpha2-macroglobulin has been identified in the hemolymph of some arthropods, but not yet from insects. Alpha2-macroglobulins are protease inhibitors, well known from mammals. These inhibitors do not work by competitively inhibiting the proteases, which is the most general mechanism of inhibiting proteases. Instead, these molecules form "cages" around protease enzymes. The cages inhibit protease activity by screening larger molecules out.

Smaller peptides are not screened out, so they can get into the active sites of the proteases, where they can be hydrolyzed. This forms a sort of selective inhibition of proteases. It is speculated that the alpha2-macroglobulins are a regular feature of arthropod hemolymph. They are thought to clear protease activity from hemolymph by collecting the proteases, then delivering them to hemocytes where they participate in a endocytotic protease clearance pathway that is similar to protease clearance pathways known in mammals.

THE BIOLOGICAL SIGNIFICANCE OF IMMUNITY

There are over two hundred species of bacteria that live in one or another association with insects. These can be categorized with repect to their pathogenicity:

- Harmless forms that are not pathogenic
- Occasional pathogens
- Obligate pathogens

Induced insect immune systems do not have antibacterial activity against all, or even, most pathogenic species. *B. theringensis* is a good example of a pathogenic species that is not troubled by immune systems. On the other hand, as far as can be known, insect immune sytems are effective against all non-pathogenic systems. What sense can we make of this? One suggestion is that insect immunity has more to do with regulating population dynamics of the non-pathogenic bacteria that naturally occur in insect bodies. My speculation is that many pathogenic bacterial species have secondarily lost protective and resistence mechanisms such that they have become labile to the induced proteins.

FACTORS OF HUMOURAL IMMUNITY

We most often regard phenyloxidases in terms of sclerotization, or tanning, of cuticle. These enzymes are found within cuticle and in hemolymph of insects. In hemolymph phenyloxidases are involved in melanin synthesis by way of quinones. Phenyloxidases are very active enzymes, and the intermediates and products of phenyloxidase reactions can be harmful to animals. These enzymes are nearly always found in a pro-phenyloxidase form. Phenyloxidases are activated in response to bacterial infections, and melanin, a tanned insoluble material is deposited around clumps of bacteria. A similar reaction occurs at the site of wounding in insect integument.

Activation of phenyloxidase has been studied in a few insects. This usually involves a signal from the surface of bacterial walls, such as the putative beta-1,3-glucan receptor. Conformational changes in this receptor activate a serine protease called prophenyloxidase activating enzyme. The activating enzyme removes a 5-kD peptide from prophenyloxidase, which yields phenyloxidase. These sort of cascade reactions are common biochemical activating mechanisms. Another important reaction cascade is the blood-clotting reaction in mammals. Lectins are agglutinating molecules that are well known from plants and many classes of animals. Lectins have been purified from the hemolymph of many insect species, including a meat fly, a cricket and a grasshopper. A possible function of some lectins may be agglutination of microbes, but the this has been clearly shown in only a single case.

HEMOCYTIC IMMUNITY

Insects defend themselves from challenges by microorganisms with a combination of humoural and cellular responses. The distinction between these two response categories is rather artificial for a couple of reasons. First, hemocytes can release some of the protein factors that follow from challenges. Hemocytes produce what we regard as humoural responses. Second, although infections elicit a complex of multifaceted responses, the responses are coordinated and integrated into a single theme: host defense. Nevertheless, it is very convenient to treat hemocytic and humoural immunity as separate topics. One point of separation is that hemocytic responses are rapid. They occur almost immediately an insect is invaded by microorganisms.

Humoural responses are much slower, they are usually not seen until some hours after infection. These longer-term responses may be more important in sustaining the hemocytic host defense reactions initiated earlier in the infection cycle.

Circulating cells are regularly found in the hemolymph of insects. These cells are known as hemocytes. Hemocytes are involved in conferring a cellular, or hemocytic, immunity to insects. These cells are formed in haemopoietic organs that are variously well developed in different insect groups. In the cricket *Gryllus* haemopoietic organs are well developed, distinct organs found in the second and third abdominal segments along either side of the heart. In other insects haemopoietic organs are amorphous clumps of cells irregularly distributed along the dorsal vessel.

You might enjoy a quote on the classification of insect hemocytes: "Classification of insect hemocytes is an issue that stimulates an instantaneous loss of good manners and an induction of bad temper at any conference on insect physiology or immunity". {Lackie, A.M., 1988, In: Advances in Insect Phyisology, Ed. by P.D. Evans and W.B. Wigglesworth, Academic Press, London} Given detectable levels of disagreement on the subject, a list of some hemocytes and their function follows:

Prohemocytes (PRs) are thought to be the stem cells from which all hemocytes arise.

Plasmatocytes (PLs) are larger cells involved in major immunological actions. PLs are capable of phagocytosis, a kind of entocytosis of smaller foreigh cells. Phagocytosis usually follows low doses of invading yeast and bacteria. PLs are also involved in encapsulation of clumps of invading microbes. There are typically more PLs than other types of hemocytes.

Granulocytes (GRs) are characterized by the presences of many intracellular granules. Electron microscope studies have revealed differences among the granules in these cells. There are structureless and structured granules. The biochemistry of the different types of granules is not yet understood. GRs are major players in encapsulation reactions.

Other hemocytes include spherulocytes, oenocytoids and adipohemocytes. These cells are regarded as "nonimmunocompetent" by some authors, although there is considerable ground to disagree with this idea. Gupta prefers to use the term "immunocytes" for GRs and PLs and "hemocytes" for all the other types. He uses these terms to stress the idea that GRs and PLs make up most of the immune responses. Spherulocytes may be involved in silk formation, melanization and secretion of some hemolymph proteins. They may also be important in hemolymph clotting responses to wounds. Oenocytoids may contribute to the final stages of cuticle formation. Other authors suggested that these cells may recognize "non-self".

DEFENSE MECHANISMS

Again, hemocytic defense responses occur immediately insects are invaded by microorganisms. GRs and PLs are responsible for a number of cellular immune reactions. They are able to phagocytize small biologic particles such as bacteria and yeast. Phagocytization refers to the process of cellular internalization of foreign invaders. Phagocytosis is not the same as killing the microbial cells. Microbes are secondarily killed after phagoccytosis.

GRs and oenocytoids are attracted to sites of wounds, where they play major roles in coagulating hemolymph to stop bleedings. We know virtually nothing about the signalling process that is responsible for moving cells to wound sites.

Another defense mechanism is nodule formation. Nodules are aggregates of hemocytes that entrap invading microbes. This is a particularly effective means of clearing the hemolymph, and it is considered to be of greater importance than phagocytosis in clearing large doses of bacteria. There appears to be a sequence of trapping by GRs, followed by aggregation by PLs. The precise sequence of events will remain a point of controversy. In addition to bacteria, viral polyhedrons, fungal spores and protozoans are taken up in nodules.

When organisms or clumps of organisms are too big to be phagocytized or to form nodules, they may be encapsulated. Encapsulation is the sequesterization of these organisms or clumps with multilayered aggregates of hemocytes. This also involves release of coagulum, in a response similar to wound-repair. The coagulum probably aides in forming layers of cells. Encapsulation was first observed by the Russian zoologist Elie Metchnikoff in 1892. He placed rose thorns into a starfish larva. The next day he saw that the thorns were completely surrounded by tiny cells. Metchnikoff called these cells phagocytes from the Greek words for eating and cells. Metchnikoff eventually moved from the University of Odessa to the Pasteur Institute, and later shared the Nobel Prize for his studies of cellular defenses in mammals and invertebrates.

We are not sure of the killing mechanisms that hemocytes elaborate. Some of the suggested mechanisms include peroxidation, melanization, and lysozymes. By comparison to our scant knowledge of invertebrate immunity, very much is understood about the killing mechanisms of mammalian host-defense cells. There are two sorts of mechanisms. One sort does not depend on oxygen. These include protease, esterase, acid-phosphatase and lysoszyme mechanisms. Another sort does depend on oxygen, including formation of singlet oxygen and superoxide anions, which are very reactive and very toxic species. Work on killing mechanisms of insect hemocytes will produce important and very novel insights into how insects protect themselves.

We recently addressed the question of how the early, hemocytic responses to infection are mediated. What we wanted to know was what biochemical signals occured between recognition of a bacterial infection and the hemocytic responses that follow. Since many aspects of host defense in mammals are mediated prostaglandins and related eicosanoids, we formed the hypothesis that these molecules are also crucial mediators of immunity in insects. The central experimental design was to inhibit various of the eicosanoid biosynthetic pathways, then observe the effects of inhibition on the insects' ability to remove bacterial infections from their hemolymph.

Since hemocytic responses occur in the first few minutes post-infection (PI), we could dissect the humoural responses from hemocytic responses by doing the experiments within the first hour PI.

In the first experiments, cells of the insect pathogenic bacterium *Serratia marcescens* were injected into 5th instar larvae of the tobacco hornworm *Manduca sexta*. Test animals were first injected with the phospholipase A2 inhibitor dexamethasone, and control animals were first treated with pure ethanol. Hemolymph samples were withdrawn at 15, 30 45 and 60 minutes PI. The samples were diluted slightly, then streaked onto agar plates. The plates were incubated for about 40 hours, then the number of red colonies on each plate were counted.

Until these data are scanned into the Web site, you'll have to look at the original publication, or use the handouts. The line connected by square boxes represents hemolymph samples from control insects: there were no red colonies on these plates, indicating that the insects had cleared all or most of the bacteria from their hemolymph. The line connected by X's represents hemolymph samples from test insects. As early as 15 minutes PI, there was a statistically significant increase in recovered bacteria. The number of recovered colonies increased over the first hour PI.

We tested the effects of several doses of dexamethasone by injecting 4 doses in different groups of insects. The results indicate that more bacterial colonies were recovered with increasing doses of dexamethasone.

The bacteria were used in these experiments are insect pathogens. We reasoned that the dexamethasone decreased the ability of the larvae to clear bacteria from their hemolymph. Since these bacteria are pathogens, the decreased clearance from hemolymph should translate into increased mortality from these infections. We tested this idea by injecting the same

4 doses that were just described into 4 groups of larvae. After 14 hours we counted the number of surviving larvae. As expected, increasing dosages of dexamethasone were associated with decreasing larval survival at 14 hours.

Dexamethasone inhibits the activity of phospholipase A2. This in turn inhibits eicosanoid biosynthesis because substrate is not available to the eicosanoid-biosynthetic pathways. We conducted experiments to test the possibility that arachidonic acid would rescue the larvae from the effects of dexamethasone. These kinds of experiments are called "end product reversal experiments", or something like that. The last bar in the figure above this paragraph shows that when larvae were treated with dexamethasone, and with arachidonic acid, larval survivorship was a great as survival in the control group.

We can see from the first figure that phospholipase A2 activity is necessary for all eicosanoid biosynthesis. We wondered if one eicosanoid biosynthetic pathway might be more important than another in mediating insect immune responses. To investigate this point we treated groups of larvae with specific inhibitors of phospholipase A2 (dexamethasone), cyclooxygenase (indomethacin), lipoxygenase (esculetin), and indomethacin and esculetin together. We also observed the effects of maleic acid, an antioxidant compound that does not interfere with the major eicosanoid biosynthesis pathways. After treating the larvae with inhibitors, we infected them with bacteria. After 2.5 hours, we withdrew hemolymph samples, and plated them on agar plates. We then counted the number of bacterial colonies on each plate. This figure shows the proportion of larvae from which ANY colonies were recovered.

This is a rigorous experiment because if any bacterial colonies were recovered, even a single colony, the individual was counted as positive. These data show that at least 1 colony was recovered from about 20 per cent of the control larvae. Colonies were recovered from about 80 per cent of the phospholipase A2-inhibited larvae. Bacterial colonies were also recovered from the cyclooxygenase-inhibited and the lipoxygenase-inhibited larvae, but from about 60 per cent, rather than the 80 per cent of phospholipase A2-inhibited larvae. These findings indicated that both pathways were involved in the immune response to infections. This idea was supported by the results from the group that was treated with both inhibitors at the same time: they were not different from the dexamethasone effects.

We concluded that products of cycloxygenase and lipoxygenase pathways were involved in mediating cellular responses to bacterial infections in this insect.

The last comment, of course, opens questions of identifying specific cellular actions that are mediated by eicosanoids. As mentioned earlier, nodule formation is responsible for clearing the bulk of bacterial infections from circulation. We developed the hypothesis that eicosanoids mediate the nodulation reaction to bacterial infections. Using a line of research similar to the one just described, we found that treating tobacco hornworms with eicosanoid biosynthesis inhibitors just prior to infecting them with bacteria resulted in severe impairment of nodulation. The influence of the inhibitors is expressed in a dose-dependent manner and can be reversed by eicosanoid-precursor polyunsaturated fatty acids. Again, we concluded that products of cycloxygenase and lipoxygenase pathways were involved in mediating nodulation responses to bacterial infections in this insect.

Finally, we began testing the broader hypothesis that eicosanoids mediate nodulation in other insect species. So far, we have tested these species:

Order	Common name
Lepidoptera	Tobacco hornworms
	True armyworms
	Black cutworms
	Silkworms
Coleoptera	King beetles - a large tenebrionid
Orthoptera	Crickets
	Cockroaches

This work serves to support the broader hypothesis that eicosanoids mediate nodulation responses in all insects that defend themselves from bacterial infections by forming nodules. Since publishing our original findings, other laboratory groups have taken up this line of work. We now know that eicosanoids mediate phagocytosis and nodulation in greater waxmoths.

Chapter 5

Transduction Mechanisms

SIGNAL TRANSDUCTION MECHANISMS

The maintenance of appropriate titres of hemolymph fuel supplies involves action of the peptide hormone, adipokinetic hormone. Problems in water and osmotic balance are also addressed by hormone action. Insects often elaborate a diuretic hormone and an antidiuretic hormone that are involved in homeostasis of salt and water balance and also in increased fluid cycling to enhance filtration.

All of these hormones, and many others in general animal physiology, are peptide hormones. Study of peptide hormones is one of the most active areas of insect physiology. One reason for this activity is that peptide hormones appear to relate to more and more areas of physiology. We also discussed the roles of prostaglandins and other eicosanoids in regulating basal fluid secretion rates in mosquito Malpighian tubules and mediating cellular immune reactions to bacterial infections. At the cellular level, the physiological actions of prostaglandins are mediated in a manner similar to the mediation of peptide hormones.

One of the most contemporary arenas of insect and mammalian endocrinology is cellular signal transduction mechanisms. Just to get a grip on terminology, we want to recognize four forms of signaling by secreted molecules. Endocrine signaling refers to the secretion of a molecule into general circulation. The secreted species will eventually interact with target cells in one or more tissues quite distant from the endocrine gland. We also recognize paracrine signaling, in which a secreted molecule is very rapidly taken into or inactivated by neighbouring cells. In this case the secreted molecule influences just a few cells in the immediate neighbourhood of the secreting cells. Paracrine signaling is related to autocoid signaling. In this case, a cell may release a molecule which interacts with receptors on the surface of the same cell. Finally, we recognize synaptic signaling, which generally takes place in nerve-nerve or nerve-muscle synapses. In all of these systems entail a general mechanism. A signal molecule is released from a cell, and the action of the signal molecule follows interaction of the molecule with a specific receptor protein.

Peptide hormones are usually very short chains of amino acids. The shortest in length is 3 amino acids in some vertebrate brain hormones (the discovery of which earned a Nobel prize for a French and an American laboratory. These two lab groups were in fierce competition to be first in learning these structures). Others are commonly a dozen or so amino acids in many insect and vertebrate peptide hormones. Insulin is a protein hormone composed of over 100 amino acids linked cysteine residues in two chains. Prostaglandins are oxygenated metabolites of certain polyunsaturated fatty acids. They feature polar substitutions on carbons 1, 9, 11 and 15. PGE_2, for example, has a carboxylic acid at C-1, at keto substitution at C-9, and hydroxyls at C-11 and C-15. Peptides, prostaglandins and a few other classes of signal molecules share the physiological feature that their actions are mediated by way of specific receptor cites located on cell surfaces.

We procede by framing two questions: [1] When any particular hormone is released in circulating hemolymph, how does it know which tissue is the appropriate target? For example, DH is released from neurosecretory cells into hemolymph, and it acts only on Malpighian tubules and on hindgut. This hormone does not act on brains or on reproductive tissues or on flight muscles. Again, how does the hormone know where to go? [2] How is a hormone signal floating around in hemolymph translated into some intracellular response? What these questions amount is wondering how a hormone signal is transduced into cellular events. The whole business is known as signal transduction mechanisms.

The first question has a rather straight-forward reply. Hormones actually "go" everywhere: they enter general circulation in very small concentrations. While in circulation they come into contact with every cell in an insect body. A general concept in endocrinology is that hormones, including locally hormone-like regulators like prostaglandins, combine with specific receptors on the surface of or are inside the cells of their target tissues. I mention inside their target tissues because most steriodal hormones, such as 20-hydroxyecdysteroid, can permeate cell membranes. Receptors for steriod hormones are usually found in the cytosol of cells. Cell surface receptors interact with peptide and other hormones and regulators that can not cross membranes. In this model one imagines that a hormone is released in general circulation and can potentially come into contact with all body tissues, but interacts physiologically only with those cells that possess receptors specific for the given hormone. Hormone receptors, often called receptor sites, are mainly large proteins. These proteins confer specificity of hormone recognition upon individual cells within tissues.

Having dispatched that issue, the next general question is, what is the biochemical basis for hormone action? This is, how is a hemolymph-borne signal translated into intracellular activity? This is a slightly more complicated area that depends upon a hormone's structure. Here, we will limit our discussion to peptide hormones and a word or two about prostagladins.

Peptides mostly adjust or regulate metabolic events. An interesting feature of peptides is that they are not able to permeate cell membranes. The hormone itself can not move from extracellular to intracellular compartments. The first issue in the physiology of peptide hormone action is that their specific receptor proteins are located on the outside of target cell membranes.

We want to make three points today:

1. This is a general sort of hormone mechanism, called a second messenger system.
2. In this system, a hormone signal, but not the hormone, gets into a cell.
3. Within cells, a low-signal hormone is rapidly amplified.

Turning attention to the figure, the transduction process begins when a blood-borne hormone interacts with a specific receptor site.

In the example, a hormone interacts with a specific receptor. This receptor is not just on the membrane surface, it is anchored deeply in the membrane. The hormone-receptor interaction causes a local change in receptor protein conformation. This conformational exposes a receptor binding site on the intracellular side of the membrane. Diffusion in the membrane bilayer eventually allows an association between the receptor binding site and a G_s protein. G_s stands for stimulatory G protein. This association stimulates an exchange in the G protein. The GDP in the unassociated G_s protein is exchanged for a GTP. The GTP causes the a subunit of the G_s protein to dissociated from the rest of the G_s complex. At the same time, a binding site on the a subunit of the G_s protein becomes available to bind with the enzyme adenylate cyclase. The a subunit binds to the adenylate cyclase, which activates the adenylate cyclase. The adenylate cyclase now converts many ATP molecules into cyclic AMP, universally expressed as cAMP.

You might think this process would go on forever. It does not. The a subunit of the G_s protein contains a GTPase activity. The GTPase hydrolyzes the GTP, which returns the a subunit to its original conformation. In its original conformation, the a subunit dissociates from the adenylate cyclase, thereby rendering the adenylate cyclase inactive. Meanwhile, the a subunit can reassociate with a b-g complex. The cycle will continue until the peptide hormone dissociates from the cell surface receptor.

ATP is converted into cAMP by release of free pyrophosphate, PPi. cAMP is a biologically ubiquitous intracellular messenger molecule. This is called a second messenger sysem beause the first messenger - the hormone - stimulates production of a second, intracellular, messenger, namely cAMP. The main point is that a message is carried across a cell membrane by changes in membrane protein conformations.

It is reasonable to ask "What does cAMP have to do with regulation of metabolic events?" The figure shows the most general reaction. There is an intracellular enzyme that goes under the name c-AMP dependent protein kinase, commonly called A-kinase. We might wonder what kinase means. This refers to an enzyme that activates another enzyme. A-kinase exists as two linked subunits, a catalytic subunit that actually performs the chemistry of activting enzymes, and a regulatory subunit that inhibits action of the catalytic subunit. cAMP acts by combining with the regulatory subunit of protein kinase. This causes a conformation change, which results in separation of the subunits, thereby yielding free, and now active, protein kinases.

The active protein kinase links a phosphate group onto another enzyme, thereby changing an inactive enzyme into an active form of the same enzyme. The phosphorylated enzyme leads to the cellular response. The figure shows several arrows that are meant to indicate a number of steps leading to the final response. Now imagine this: inactive phosphorylase kinase becomes active phosphorylase kinase after A-kinase links a phosphate group onto it. Now let's ask what does kinase do? The kinase hooks a phosphate group onto still another enzyme, thereby changing it from inactive to an active form. In the example, the phosphorylase kinase activates glycogen phosphorylase. Looks like a whole lot of activating going on. The glycogen phosphorylase produces the final cell response to the hormone: release of free sugar from glycogen.

Why are there so many steps in this process? The answer lies in realization that this is correctly called an amplification cascade, which tells us how the incoming signal can be amplified to produce peak cellular activity level in seconds. The key is that each step in the cascade is catalytic. By way of example, imagine that each step in a cascade operates only ten times. Ten is a completely arbitrary number I selected to make the point of amplification:

- Step 1 gives 10 molecules of cAMP
- Step 2 gives 100 active protein kinase molecules
- Step 3 gives 1,000 active kinase molecules
- Step 4 gives 10,000 enzyme molecules
- Step 5 gives 100,000 metabolic steps

Again, these numers are not data. I made them up to illustrate rapid amplification of a hormone signal.

Now, the next question to ask is "How do we turn this cascade off?" cAMP production depends upon continued secretion of hormone. Once hormone secretion stops, the hormone-receptor complex dissociates, usually with deactivating metabolism of the hormone. All cells capable of making cAMP express an enzyme called phosphodiesterase. This is a continuously running enzyme that rapidly hydrolyzes any cAMP it can find. When cAMP formation is stopped, intracellular cAMP levels rapidly approach zero. Then protein kinase subunits reassemble, whereupon the phosphorylated enzymes are dephosphorylated to inactive forms by another set of enzymes called phosphatases. Cells are thereby returned to their unstimulated state, tiny enzymic-sized beers are opened and things sort of chill out until the next episode.

To summarize, a hormone interacts with a specific receptor cite, and the hormone-receptor interaction leads to increased intracellular cAMP concentrations by activation of an effector protein, in this case, adenylate cyclase. There are other effector proteins. The receptors are coupled to the effector proteins through intermediates known as G proteins.

It has been my experience that once this system is revealed to people they generally change majors to physiology and biochemistry.

SIGNAL TRANSDUCTION MECHANISMS II (CALCIUM)

A hormone signal could be transduced into increased intracellular cAMP concentrations, which in turn activates enzymes involved in some cellular responses to the hormone. The main idea is recognizing that a signal, rather than a hormone *per se*, leads to a cellular response. cAMP is a universal intracellular messenger. Changing concentrations of cAMP are translated into changes in enzyme activities, and the enzyme activities produce the cellular response.

Most cells express multiple signal transduction pathways. In addition to the cAMP pathway, nearly every cell expresses a Ca^{++} signal transduction pathway. As we saw for cAMP, Ca^{++} is another universal intracellular messenger. The cellular actions of these two intracellular signals are fairly similar. Both messengers act by binding to specific proteins, and the binding activates the proteins by causing them to undergo conformational changes. Very often, multiple intracellular messengers interact with each other, and the cross-talk among signal transduction pathways influences the intensity of a cellular response to hormones or other extracellular signals. For an example, a diuretic hormone known as AVP increases fluid secretion in mammalian kidneys. At the same time, certain prostaglandins attenuate the intensity of fluid secretion. Hence, the overall secretory response to the hormone is the outcome of multiple signal transduction pathways.

An important condition must be met to allow Ca^{++} to serve as an intracellular second messenger. Ca^{++} must be present in very low concentrations, typically something like 10^{-7} M, in the cytosolic fraction of cells. Most eucaryotic cells elaborate a sophisticated system of mechanisms to maintain these low

Ca^{++} concentrations. Most cells have a Ca^{++}-ATPase in the plasma membrane, and this enzyme serves as a Ca^{++} pump, moving Ca^{++} ions against a chemical gradient from the cytosol to the extracellular spaces at the expense of ATP. The ions are pumped against a gradient because Ca^{++} is present in substantial concentrations, around 10^{-3}M, in extracellular spaces. Some cells, such as muscle cells, have an additional Ca^{++} pump in their plasma membranes.

This is known as a Na^{+}-driven Ca^{++} antiport, or exchanger. This antiport works by coupling the inward transport of Na^{++} with the efflux of Ca^{++}. The antiport is particularly important when Ca^{++} concentrations are elevated by frequent stimulation of muscle cells. Both pumps maintain low intracellular Ca^{++} concentrations by pumping Ca^{++} ions into the extracellular spaces.

Whereas the Ca^{++} concentrations in the cytosolic compartment of cells are very low, there is a substantial amount of Ca^{++} in eucaryotic cells. Most of the Ca^{++} is segregated in a specialized compartment called a calcium-sequestering compartment. The membranes of this intracellular compartment also have a Ca^{++}-ATPase that pumps Ca^{++} ions into the compartment. The plasma membranes and the membranes of the calcium-sequestering compartment have Ca^{++} channels that are briefly opened by transduction of extracellular signals. Ca^{++} ions rapidly move down their chemical gradients, into the cytosolic compartment, creating a dramatic, local increase in Ca^{++} concentrations. The brief spike in intracellular Ca^{++} activates proteins in the Ca^{++} signal transduction pathways.

Let us turn, now, to the mechanism of releasing Ca^{++} from the intracellular compartment. The mechanism involves phospholipid metabolism. The plasma membranes of eucaryotic cells are comprised of lipids, mainly phospholipids, and proteins. Here is a typical cell membrane, showing a bilayer of phospholipids. With respect to bulk quantities, phosphatidylcholine and phosphatidylethanolamine are the main components of biomembranes. Minor amounts of many other lipids are also present in membranes. One of these quantitatively minor lipids is known as phosphatidylinositol (PI). PI is just a garden-variety, off-the-shelf phospholipid with a single molecule of a sugar, specifically inositol, in the polar head group. PI usually makes up less than 10 per cent of cell membrane lipids.

Three forms of PI are present in biomembranes. An enzyme known as PI-kinase can add a phosphate to carbon-4 of the inositol moiety, creating phosphatidylinositol 4-phosphate (PIP). Then another enzyme, known as PIP-kinase can add a phosphate to carbon-5 of the inositol moiety, creating phosphatidylinositol 4,5-bisphosphate (PIP_2). PI, PIP and PIP_2 are the three forms of PI in cell membranes. PIP_2 is present in very low amounts, typically less than 10 per cent of all PI forms, and less than 1 per cent of membrane lipids. As usual in cell signaling, despite its low concentration, the PIP_2 is the most important part of our story.

Compared to our understanding of the cAMP pathway, the details of the Ca^{++} signaling pathway are not as well understood. It is thought that the Ca^{++} signal transduction mechanism begins with interaction of an extracellular signal with a cell surface receptor. As we saw in the cAMP pathway, the receptor undergoes a change in conformation, and interacts with a G-protein, known as the Gp protein. The Gp protein then activates an enzyme known as phospholipase C (PLC). The particular PLC is is specific to PIs. This enzyme cleaves PIP_2 into two products, a diacylglycerol (DAG) and inositol triphosphate (IP_3). The inositol has three phosphates, two from the PI kinase and PIP kinase actions on carbons 4 and 5, and the third from the phosphodiester linkage to the glycerol backbone of the PI.

IP_3 is a water soluble, small molecule. It is biologically inactivated by action of a specific phosphatase that removes one phosphate. As an active messenger, IP_3 interacts with Ca^{++} channels in the calcium-sequestering compartment. This interaction allows a rapid and brief increase in cytosolic Ca^{++} concentrations. The Ca^{++} ions activate target proteins, and thereby translate an extracellular hormone signal into a specific cellular response.

Ca^{++} does not directly activate target proteins in cells. Free Ca^{++} ions are bound by a very specific Ca^{++} binding protein known as calmodulin. Calmodulin in a ubiquitous protein, found in all eucaryotic cells analysed for it. As we have seen before, Ca^{++} binding induces a conformational change in calmodulin. The Ca^{++}-calmodulin complex is now able to activate proteins by binding to the target. The target protein undergoes a conformational change and becomes activated. This is represented in a figure.

The overall physiology can be summarized like this: an extracellular messenger binds to a specific receptor on a cell surface. The receptor changes conformation, and interacts with a Gp protein. The Gp protein activates a PLC, which splits PIP_2 into DAG and IP_3. The IP_3 interacts with Ca^{++} channels on the membrane of the calcium-sequestering compartment. Ca^{++} flows into the cytosolic compartment where it interacts with calmodulin. The Ca^{++}-calmodulin complex then activates target proteins.

The story is not quite complete because the PLC step actually activates two signaling pathways. The DAG formed by hydrolysis of the IP_3 may express two roles in cell signaling. For one, arachidonic acid may be hydrolyzed from the DAG. The free arachidonic acid is taken into biosynthesis of prostaglandins and other eicosanoids. As we mentioned before, eicosanoids can modulate cellular responses to hormones. For just a bit more detail on eicosanoid actions, most eicosanoids are thought to act through specific cell surface receptors. The receptors act through G proteins to modify the cAMP and C-kinase pathways.

DAG activates a separate pathway known as the Protein kinase C, or C-kinase, pathway. Protein kinase C binds DAG, which causes a slight conformational change. In its changes form, protein kinase C expresses increased affinity for Ca^{++}, which then binds to protein kinase C. When bound to DAG and Ca^{++}, the active protein kinase C transfers a phosphate from ATP to specific serene or threonine components of target proteins. A common example is the phosphorylation and activation of the plasma membrane Na^+-H^+ exchanger. This exchanger regulates intracellular pH. For an example closer to our interest, AKH also acts through specific cell surface receptors in fat body of the migratory locust. AKH stimulates increased intracellular cAMP, release of IP_3, and increased intracellular Ca^{++} concentrations. These changes in intracellular messengers are responsible for release of sugar residues from cellular glycogen stores and for mobilization of fat body lipids.

The role of IP_3 in releasing Ca^{++} from intracellular stores, and the role of Ca^{++} as an intracellular second messenger are basic biological actions in cellular physiology. Michael Berridge, now a leading administrator in the Medical Research Council in England, is credited with discovering the importance of the IP_3 signal transduction pathway. Used the sheep blowfly salivary gland as his research model. The isolated salivary gland secretes saliva in response to 5-hydroxytryptamine, another biogenic amine. Fluid secretion in the isolated glands is mediated by increased intracellular cAMP concentrations and by the IP_3

INSECT MATINGSYSTEMS

BIOCHEMICAL MECHANISM OF RELEASING EGG-LAYING BEHAVIOUR

Our discussion of insect reproduction seemed to assume female insects were sort of automatically endowed with sperm at the correct times in their lives. This is not always a correct assumption. Indeed, a certain proportion of adults of most insect species fail to locate mates. The business of locating mates, courtship interactions, mating, fertilization, depositing eggs, and caring for and provisioning offspring (or not doing these activities) are aspects of mating systems. Many of these components of mating systems are mediated through neuroendocrinological mechanisms. The idea of neuroendocrine systems draws attention to the physiological basis of insect mating systems. A very common example comes from considering the roles and actions of sex pheromones in the mate-seeking phases of mating systems.

Lurking behind these components, or phases, of mating systems is still another aspect of insect reproduction. Males have another evolutionary issue. Beyond locating, courting, and mating females, males are constrained by a need to assure the paternity of self. Strong selective pressures seem to favour mechanisms that assure an individual's paternity. I would just speculate that individuals lacking paternity assurance mechanisms would probably not be well represented in succeeding generations. Insects express many behavioural mechanisms, such as mate guarding, to assure an individual's paternity. Insect also elaborate a number of biochemical and physiological mechanisms to enhance an individual's chances of assuring paternity. Males and females seem to cooperate in some of these mechanisms.

The close connection between physiology and behaviour in insect systems. We will consider a mating system because I have a fair bit of experience in this work, and because there sufficient information to present something close to a detailed picture of the system. The mating system of the Austrialian field cricket, *Teleogryllus commodus*, features behavioural and biochemical mechanisms of assuring paternity.

TEMPORAL ORGANIZATION OF A MATING SYSTEM

Nearly everybody knows that male crickets attract their potential mates by emitting a calling song. Singing is an expensive business for male crickets. Males often maintain and defend a territory that, will later serve as a courtship and mating arena. Male crickets produce their songs by stridulation of the their wings. The singing itself is an energy-intensive process. Before they begin singing, males of some crickets create a small burrow. The burrow acts as a sort of amphitheater, extending the males' broadcast distance. Does the song of a conspecific male sound like some kind of recording star? Maybe a Bob Dylan or a Pavarotti?

We begin with the calling song because the temporal organization of male and female mating behaviours increases the chances of successful mate location for both sexes. First, calling behaviour is under circadian rhythm control. Males express their calling behaviour on a daily basis in a predictable rhythm. Of course, females are attracted to the calling song. Males must be prepared to court and mate with a female by the time he begins his singing.

There are physiological and behavioural preparations. Again, many males prepare a burrow. There is more. As you already know, most (but not all) cricket species mate by transferring a spermatophore from male to female during copulation. Males of *T. commodus* synthesize a new spermatophore each day. As seen for singing, spermatophore synthesis is also under circadian control. In the cricket I worked on, spermatophore synthesis takes place in late afternoon, before nightly singing. The point to take is that two aspects of the mating system are expressed in a regulated temporal organization.

Female crickets also exhibit a number of sexual behaviours. One is her nightly locomotor behaviour. We can show this in an actogram. The top bar shows a 12:12 hour photoperiod. We take three points. First, virgin females express a nightly locomotor behaviour, that is, they go out walking around the park. We might presume they walk around listening for calling songs. The temporal organization of the female walking happens to match up with the male singing rhythm. Hmmm..

The second point is shown in the column under 'S'. The call song can stimulate virgin females into locomotor activity is the song is played during normally inactive periods. I guess this helps out in cases where the clocks are set just right. The third point is that after mating, the female is no longer responsive to calling songs, and she no longer expresses here nightly locomotor activity. The main point here is the sexual berhavior of males and females is characterized by a temporal organization that facilitates mate finding.

As just mentioned, females change their sexual behaviours after mating. As mentioned, females are no longer stimulated to walk by calling songs, and the nightly locomotor behaviour ceases. We might think these two points make sense because a mated female no longer needs to risk her life walking around in dangerous parks. These behavioural changes also facilitate paternity assurance because the a mated female is more likely to remain in the male's territory. Females also exhibit a third behaviour right after mating. There is a very large increase in egg-laying behaviour, usually in the first 24 hours following mating.

This is shown in a figure. We have a three-dimensional plot of egg-laying behaviour. The x-axis indicates individual females. The y-axis indicates numbers of eggs deposited by each individual, and the z-axis shows four time points. The first x-axis represents egg-laying behaviour in the 24-hour period before mating. This shows that sexually mature virgin females will typically deposit a small number of sterile eggs. The next axis represent the first 24 hours after mating, during which most individuals have deposited a vastly increased number of eggs. The next line shows some females will wait for another 24 hours before laying her eggs. I guess this occurs more rarely than suggested by these data, because the nearly all the females I have experienced will deposit their eggs during the first night.

The burst of egg-laying activity has been called an "egg-flood" because so many eggs are deposited. This is a misleading term because each egg is deposited in a controlled behavioural programme. The eggs do not "flood" out of the females. The main points here relate to females. The temporal organization of female sexual behaviour facilitates mate finding, and mating radically alters female behaviours.

These changes in female sexual behaviour, reduced response to male calling songs, reduced locomoter activity and a sustained programme of egg-laying, are cooperative. Together, these changes form the characteristic post-mating behavioural syndrome. The physiological basis of changing the first two behaviours remains unknown. On the other hand, the mechanism of releasing oviposition behaviours is better understood. We know that the release of egg-laying behaviour is not physiologically coupled to the other behaviours.

Let us consider the anatomy of cricket female reproductive tracts. In this drawing one ovary is absent to show structures that otherwise would be obscured. OP indicates the ovipositor, GC the genital chamber, ACG the accessory glands, M the oviducal muscles, LOD the lateral oviduct, LAG the last abdominal ganglion, SPD the spermathecal duct, SP the spermatheca and OV the ovary. A successful mating entails placing a spermatophore into the genital chamber. A fine spermatophore tube makes its way through a pore in the ceiling of the genital chamber, and up into the spermatheca duct. This step is necessary for complete transfer of the seminal fluids into the spermatheca. A fair proportion of matings fail on this point, and examination of spermathecae from these failed matings shows the seminal fluids did not reach the spermatheca.

Aside from sperm, the enzymes and substrate required for biosynthesis of PGE2 are transferred along with the seminal fluids. Once the seminal fluids reach to spermatheca, the transferred enzyme activity converts arachidonic acid into PGE2. The PGE2 moves out from the spermatheca and into general circulation. I believe, with knowing for sure, the circulating PGE2 releases egg-laying behaviour by interacting with specific receptors in the last abdominal ganglion. This is known as the enzyme transfer model of releasing egg-laying behaviour. The model is pretty well supported, although we do not have data on the last point.

Here is a summary of the evidence on the enzyme transfer model.

- PGE2 releases egg-laying behaviour in sexually mature virgin females
- Spermathecae from mated females contain about 500 ng PGE2/spermatheca. Those from virgins had no detectable PGE2

- Spermathecae from mated females synthesized PGE2 at substantial rates, over 30 pmol/hr; those from virgins synthesized none
- Spermatophore contents synthesized PGE2 at about the same rates seen in females, 25 pmol/hr. The spermatophores contained almost zero PGE2

On the first point, when PGE2 was injected into the hemocoel of sexually mature virgins, the females responded by releasing egg-laying behaviour during the following 12 hours. Second, PGE2 could be detected in spermathecae from mated, but not virgin females. Spermatophores themselves contained only very small amounts of PGE2, indicating that PGE2, per se, is not transferred to females during mating. third, PGE2 biosynthetic activity was detected in spermathecae from mated females, and in spermatophores, but not in spermathecae from virgin females. Taken together, these data strongly supported the enzyme transfer model.

The next question to ask relates to the specificity of PGE2 in releasing egg-laying behaviour. Do all PGs release the behaviour, or is the action specific to E-type PGs? We understand from our consideration of receptor sites and signal transduction mechanisms this is an important question. If the PGs act in a receptor-mediated physiology, then we would expect to observe specificity to E-type PGs. Let us consider structures of PGs. All 1-series PGs are formed from 20:3n-6. The 2-series PGs are formed from 20:4n-6, and the 3-series PGs are made from 20:5n-3. PGs of each series have 2 fewer double bonds than their parental fatty acids. The key point is most PGs look pretty much alike. They are all 20-carbon carboxylic acids with substitutions on C-15 and a 5-membered ring. The substitutions on the ring define the letter series of each PG.

To investigate the relationship between PG structure and egg-laying activity, we injected a standard dose of PG into the hemocoel of sexually mature virgins, then observed the egg laying responses. We worked with 16 different compounds. We arranged these structures into three groups, based on structural features. The compounds in group 3 are not the standard PGs. We tested the egg-laying activities of a monounsaturated fatty acid, to control for a general pharmacological effect of lipids. We used a PG that is sterically indered by an oxygen bridge, and a lipoxygenase product that lacks the 5-membered ring of PGs. None of these compounds release egg-laying behaviours.

At the other extreme, highest egg-laying activity was associated with the group 1 PGs, all variations on PGE2. All feature a 5-membered ring, and in every case the ring is substituted with a C-9 keto and a C-11 hydroxyl. These particular ring features are important. We can compare the egg-laying activity of other PGs lacking this arrangement, those in group 2. The group 2 PGs lack the PGE features, and these yielded zero to only low egg-laying activity. Overall, there is a biological specificity for E-series PGs. Again, this finding is consistent with the idea that the PG action is mediated through receptor sites. The receptor sites can discriminate among closely related molecules.

The enzyme transfer model indicates the enzyme activity from males converts arachidonic acid into PGE2 within the spermathecae of newly mated females. The PGE2 subsequently moves into circulation, then releases the behavioural programme of egg-laying. One of the most direct tests of this idea would be to place radioactive arachidonic acid into spermathecae of mated females, then see if radioactive PGs turn up in their hemolymph. The practical problem is it is next to impossible to inject anything into a spermatheca. After trying this a few times, I conceived the idea of letting a male cricket inject radioactive arachidonic acid into spermathecae.

We started this by injecting about 20 mCi of radioactive arachidonic acid into males on the day they emerged as adults. These males started producing a spermathophore each afternoon about 3 to 5 days later, according to their usual development. Here is where it can be quite useful to know males produce their spermatophores on a circadian rhythm. I collected spermatophores each afternoon, then assayed them for radioactivity. The males produced radioactive spermatophores for the next 49 days. So males are able to produce radioactive spermatophores.

I wanted to determine the radioactivity was associated with arachidonic acid, and not some uncharacterized metabolic by-product. I collected a bunch of spermatophores, then extracted total lipids from them. After isolating phospholipids on thin layer chromatography, we prepared methyl esters of the fatty acids. We analysed these by radio-high performance liquid chromatography. There was a single peak of radioactivity corresponding exactly with authentic arachidonic acid. So the males somehow packaged radioactive arachidonic acid into the phospholipid fractions of their spermatophores.

This work set the stage for the experiment I wanted to perform. I was interested in seeing the actual sexual transfer of arachidonic acid from male to female. To look at this I allowed my pre-labeled males to court and mate with untreated virgin females. It usually takes about an hour for the contents of a spermatophore to ooze into a spermatheca. One hour after mating with a pre-labeled male, I isolated the spermatheca from each female. The spermathecae from these matings contained about as much radioactivity as a typical radioactive spermathophore. On the other hand, about 24 hours after mating with pre-labeled males, the spermathecae contained no radioactivity. We see the males can transfer radioactive arachidonic acid to females.

The overall data are consistent with transfer of radioactive arachidonic acid from male to female, followed by movement of the radioactivity from the spermatheca into the hemolymph. Of course, the next experiment was to collect hemolymph from females mated with pre-labeled males, and characterize the radioactivity in the hemolymph. Most of the radioactivity in female hemolymph was associated with PGE2 or polar products of PGE2 metabolism. These findings indicate the PGE2 in the hemolymph of mated females is derived from arachidonic acid from males.

Now, this work was done on a single cricket species, *Teleogryllus commodus*. Overall, the work supported the enzyme transfer model, now amended to indicate males transfers and enzyme and substrate for PG biosynthesis in females. We might think this sort of mechanism occurs in many insect species, but it is not so. PG levels are greatly increased in the female reproductive tracts of many insect species. These include:

- House cricket
- Silk moth
- Cabbage looper
- Migratory locust
- Housefly

The mechanisms of increasing PG levels differ among species. As you might expect, the mechanism in the house cricket and the locust is just like the Australian field cricket mechanism. The moths transfer preformed PGs, per se, in their matings. Males of the housefly mobilize and transfer arachidonic acid, but not the enzyme, during their matings. In all cases, PGs increase in female reproductive tracts following mating.

Do these PG increases always release egg-laying? The answer is certainly not. This has been tested in quite a few species, and so far only crickets and the silk moth seem to positively release egg-laying in response to PGs. We conclude that PGs are involved in other, still unidentified, aspects of insect reproductive biology.

Let us return to our theme of assuring paternity, by asking what is the biological significance of the cricket mating system? First, the reproductive success of an individual male is related to how many of his sperm are successful in fertilizing eggs, not to how many females acquire his sperm. By casting the issue in terms of sperm success, rather than mating success, we introduce an area of evolutionary biology known as sperm competition. Mechanisms to avoid or reduce sperm competition are recognized as a powerful force in the evolutionary shaping of mating systems.

Two conditions must obtain to allow the occurrence of sperm competition. One, females must be able to mate with more than one male. Two, sperm must remain viable in the body of the female. Sperm competition is not at all restricted to insect biology. There are many examples of the influence of sperm competition in other invertebrates as well as vertebrate mating systems.

One of the basic experimental strategies in assessing sperm competition is known as multiple mating experiments. In these experiments a single female is allowed to sequentially mate with two or more males. Among insects, the general finding is that the first male does not always fare quite so well. In many experiments the last male's sperm fertilizes 75 to 100 per cent of the eggs. Here are two examples of strategies to increase chances of the mating success of the last male. Fruit flies displace sperm from previous mating. Similarly, locusts "flush" sperm from previous mating. These actions serve to reduce the competition for egg fertilization posed by the sperm populations already in a female's possession. There are many adaptations, behavioural and physiological, that appear to give a competitive edge to one's own sperm, relative to another population of sperm. Some of these adaptations seem to work in concert. The cricket mating system includes a biochemical mechanism to release egg-laying behaviour. For another component of the same system, males perform their calling, courtship and mating in a pre-constructed arena. Due to her own syndrome of post-mating behaviours, the females tend to remain in their arenas. For still another component of the system, males tend to guard their newly mated females until the eggs are deposited.

SOME MECHANISMS THAT APPEAR TO REDUCE SPERM COMPETITION

Plugs

These serve as physically barriers, blocking the entry of additional sperm populations. These are seen in various insects, spiders, snakes and mammals.

Prolonged Copulation

The male himself acts as a sort of plug. This is seen in many insects and in certain vertebrates.

Post-mating Guarding

Seen in many insects, including dragonflies, dungflies and in the cricket system.

Take-over Avoidance

Includes going to places of low risk, and maintaining territories.

Transfer Chemical agents Altering Behaviour

These agents might suppress female receptivity, or release oviposition prior to subsequent mating.

ENDOCRINOLOGY OF INSECT DEVELOPMENT

The events that take place in the egg stage of insect life cycles are known as embryonic development. Upon emerging from eggs, juvenile insects embark on an excursion of post-embryonic development that will eventually take them to their adult forms. Post-embryonic development is characterized by dramatic increases in size and by formation of new anatomical structures characteristic of adults. Argued from an ecological point of view, juvenile stages of insects are devoted to eating, growing, and accumulating energy reserves while adult stages are devoted to migrations, mate-finding and reproduction.

This appears to be particularly true among some insects, such as certain species of sphinx moths, which do not feed during adulthood. But as generalizations, these broadly sweeping remarks are not strictly true. Development of reproductive organs begins early in larval life in many insect species, so even if an insect is not actually reproducing, it is preparing for reproduction quite early in life. Many adult insects spend quite a lot of time in feeding behaviours. The value of broad generalizations is that we can appreciate the major thrust of insect life stages. Certainly larval insects are largely given over to eating and growing.

Insects are invested in a more or less rigid exoskeleton, and a generally accepted wisdom is the physical constraints of a least certain of the very rigid parts of exoskeletons inhibit growth beyond set limits. Here is an example of the overall increase in weight during several weeks of development. The figure emphasizes the potential to increase in size during post-embryonic development, in this species from a few milligrams at hatching to nearly 2 grams.

Fully hardened cuticle does not expand, so growth of the hardest parts can only take place during molts. The softer parts of insects can expand. The softer cuticle often occurs in folds that can be expanded, and sometimes the cuticle itself can be stretched. Another figure indicates that the soft parts of an insect grow rather continuously, while the hard parts grow in steps.

Growth can refer to an increase in cell number or an increase in cell size. The development and expression of cuticle depends on the cells of the epidermis. Epidermal cell numbers increase prior to molting in many insect species. In larvae of many Diptera, growth is due exclusively to increased size of the epidermal cells. Similarly, growth of internal organs can proceed by increased cell size or increased cell number. Both forms of growth occur in the mosquito *Aedes*. Cell numbers in the nervous system and fat body increase, but most other tissues have a constant number of cells. Growth occurs by cell enlargement in these tissues.

Malpighian tubules are separated into primary Malpighian tubules and secondary Malpighian tubules. The primary tubules arise from cells of the proctodeum during embryonic development. The secondary tubules emerge during post-embryonic development. The secondary tubules are reduced to buds during each molt, then develop by cell division, following by cell enlargement.

Let us turn to the endocrinology of molting. We will consider five hormones in the regulation of postembryonic development, plus a newly discovered, sixth hormone. The actions of the five hormones are summarized in a figure.

Prothoracicotropic hormone (PTTH) is synthesized in neurosecretory cells of the brain. It is stored and released from the corpora cardiaca. Again, PTTH is released from the neurohemal portion of the CC. PTTH stimulates the prothoracic glands to release ecdysone into the hemolymph. The prothoracic glands have other names. You will see the prothoracic glands called by other names, including ventral glands and ecdysial glands. In higher Diptera it is part of the ring gland. Ecdysone is not the active molting hormone. Various tissues, including the fat body convert ecdysone to 20-hydroxyecdysone, the active form of molting hormone, shown here.

The cells of the epidermis respond to 20-hydroxyecdysone with initiation of the process of molting. Apolysis refers to the detachment of old cuticle from the epidermis, the first step in molting. Molting fluid is actively transported into the space between the old cuticle and the epidermis. Epidermal growth by cell division may occur during this time. The next step involves depositing layer of new cuticle. This is followed by activation of enzymes in the molting fluid that function to digest inner layers of the old cuticle.

The amount of juvenile hormone (JH) released from the corpora allata (CA) determines the form of the new cuticle that is deposited. When JH is present in high concentrations, the new cuticle is larval; when JH is present in low concentrations, the new cuticle is pupal. Adult cuticle is formed in the absence of JH.

Sir Vincent Wigglesworth, who was knighted for his research in medically important insects, first suggested the idea of an "inhibitory hormone" in the 1930's. Working with penultimate instars of the blood-sucking bug, *Rodnius*, Wigglesworth showed that if bugs were decapitated 5 days after a blood meal, they molted into the last larval instar. If the bugs were decapitated 3 days after a blood meal, they molted into precocious, but miniature adults. The Harvard biologist Carrol Williams gave JH its name. Williams worked with *Cecropia* moths. The adult moths were parabiotically joined to diapausing pupae. Parabiotically is a relatively rare word for people who work on mammalian systems.

Parabiotic animals are joined in such a way that they share hemolymph circulation. Wigglesworth was famous for his "parabiotic trains", in which many *Rhodnius* nymphs were joined in long trains; molting would progress rearward down the trains. Williams' idea was to try to prolong the life span of the adult moths, which do not feed, by providing nutrients from the pupae. The interesting thing was this: when the pupae were joined to adult females, the pupae would continue normal development and emerge as adults. Alternatively, when the pupae happened to be joined to male adults, they molted into another PUPAL stage.

A series of now famous experiments finally showed that males of some moths, including *Cecropia* moths, accumulate a blood-borne factor in their abdomens that exerts a "status quo" effect on pupal moths. Williams named the factor juvenile hormone because it caused molting into a second pupal stage.

Ecdysis refers to shedding the old cuticle, and is the basis of the name ecdysone for molting hormone. We note that apolysis and ecdysis are separate processes. In some insects shedding of the pupal skin by adults is triggered by another brain hormone called eclosion hormone (EH). EH is synthesized in the median neurosecretory cells of the brain, and stored in the neurohemal portion of the CC. It is released from the CC under regulation of a circadian rhythm. EH stimulates the abdominal nervous system to play out a "hard-wired" behavioural programme that helps the adults emerge from the pupal case.

We consider a figure, showing the influence of EH on nerve firing in an in vitro preparation of the tobacco hornworm ventral nerve cord. EH was added to the preparation 22 minutes before the first spike occurred. The early spikes (A and B) represent the rotational bursts that begin ecdysis. The later spikes (C and D) represent peristaltic bursts. Jim Truman and Lynn Riddiford, then at Harvard, are responsible for the discovery of EH, and its role in helping insects emerge from the pupal case. Truman and Riddiford married, and took up dual positions in the Department of Zoology at the University of Washington in Seattle.

Let us now introduce a newly discovered hormone in the physiology of insect development. While molting appears to be rather abrupt, the process is long and detailed. As molting progresses, the inner layers of the old cuticle are digested away. Then a soft new cuticle is deposited. When these developmental events are complete, the shedding processes begin, and the new cuticle is expanded and hardened. The key point is noting the shedding processes are behavioural, and the role of EH is to release a pre-programmed sequence of behaviours. We now know the release of the ecdysial behaviours is more complicated.

Last year Zitnan and colleagues, then at UC Riverside, discovered the existence of a second ecdysial hormone, called Mas-ETH. Mas comes from *Manduca sexta*, and ETH stands for ecdysis-triggering hormone. ETH is a 26 amino acid peptide. It is produced by a system of epitracheal glands. In *M. sexta* there are 8 pairs of epitracheal glands, each located on the outer surface of the large tracheal tube immediately adjacent to each spiracle. The hormone is produced and released from Inka cells. Inka is the name of a fairy goddess, known from the Tatra Mountains. Zitnan takes great inspiration from Inka, and he named the cells in her honour.

Ecdysis is now thought to follow from the sequential release of two hormones. EH is released from the brain, and it primes the Inka cells to release ETH. The Inka cells release ETH, and the ETH acts on the abdominal ganglia to release the behaviours required for ecdysis. But why do we need two hormones to release ecdysial behaviours? Ecdysis involves many behaviours that are unique to ecdysis. Successful ecdysis requires that all behaviours are coordinated. Failures in coordination result in deformities and death: each ecdysis episode is a crucial event in an insect's life.

The Inka cells are positioned near each spiracle, where tracheal linings are withdrawn at ecdysis. These linings are fragile, and if torn the tracheal supply to that region is blocked. It is thought the Inka cells are in best position to serve as a last inspection site before the insect is irrevocably committed to going through ecdysis.

The physiology of molting involves apolysis, depositing new cuticle and ecdysis. The circulating titre of JH at the time of the molt determines the developmental fate of the insect. Here is an example of the relationship between ecdysone and JH through 4th and 5th larval instars, the pre-pupal wandering stage and pupal stage of the tobacco hornworm. The solid line shows ecdysone titres.

Ecdysone titre spikes a day or so before molting into the 5th instar, then again just before the wandering stage as a prepupa. There is a small increase in ecdysone titre before the pupal molt. Ecdysone titres reach a higher level and remain elevated during the pupal stage. The dotted line represents JH titres. There are high titres of JH in the 4th instar, which are lower in the 5th instar. A small spike of JH is associated with the wandering phase, and no JH is associated with the pupal stage.

JH is also active in adult stages of insect life. JH plays an important role of regulation of vitellogenin synthesis in adult female fat bodies. JH acts directly upon follicular epithelia of ovaries to facilitate uptake of yolk proteins. JH is also active in adult males, where it regulates development of reproductive tract accessory glands.

Hormones are transported in circulating hemolymph. Since the hemolymph bathes virtually every cell in the insect body, circulating hormones have the potential to come into contact with all tissues. Again, specific receptor sites form the connection between a circulating hormone and particular target cell. JH and ecdysteroids are not water soluble, and they are associated with specific high-affinity transport proteins. These transport proteins, then, are an important link in the overall physiology of endocrine regulation of biochemical and physiological events. These proteins are also partly responsible for regulating circulating hormone levels. They can act by protecting the hormone from enzyme degradation and by preventing rapid loss of hormone from the hemolymph through excretion.

The titre or levels of circulating hormones are regulated at several points. Biosynthesis, release, transport and degradation of hormones are regulated events. Biosynthesis of ecdysone depends upon PTTH from the brain. JH biosynthesis is partly regulated by nervous signals from the brain to the CA. Ecdysone is degraded to a biologically inactive form, 3-dehydro-alpha-ecdysone, and 20-hydroxyecdysone may similarly be oxidized to a 3-dehydro form.

There are two major pathways that change JH into biologically inactive forms. The minor pathway is hydration of the epoxide functionality to the diol. The most important mechanism is hydrolysis of the JH ester to its respective acid. JH can be broken down by general carboxyesterases, enzymes that occur throughout the insect body and by JH-specific esterases. The JH-carrier protein protects circulating JH from the general esterases, but not from the JH-specific esterase.

The endocrinology of insect reproduction is not without controversy. Mosquitoes elaborate an egg development neurohormone (EDNH) that is produced in the brain. EDNH is released after females take a blood meal. The hormone acts on the pre-vitellogenic ovary, and stimulates release of ecdysone, which is converted to 20-hydroxyecdysone. The 20-hydroxyecdysone then acts on the fat body, which results in vitellogenin synthesis, and development of the vitellogenic ovary. The interpretation of these data is that in mosquitoes the ecdysteroid produced by the mosquito ovary acts on the fat body to stimulate synthesis of yolk proteins. This, of course, runs counter to the conventional wisdom that JH is released from the CA, then acts on the fat body to stimulate yolk protein synthesis. Among people who are interested in the topic, there are two camps: one group absolutely disagrees with this idea.

Other reproductive hormone actions have been postulated for some insect species. Ken Davey and his co-workers in Canada suggested that there is an internal feedback control of ovarian development in the bug Rodnius. In this insect an antigonadotropin is thought to inhibit the action of JH on ovarian follicles to prevent from developing patency. This is regarded as a form of direct ovarian inhibition. Indirect ovarian inhibition is thought to occur in several species of Diptera. The general scheme is that the follicle cells produce an oostatic hormone that acts on the neuroendocrine system, causing a reduction of JH.

Chapter 6

Terrestrial Insects

We can regard the current line of talks as a discussion of physiological adaptations to terrestrialization. Perhaps you can try to integrate several areas of physiology along those lines, including information about the integument, the respiratory system and the circulatory system. In our last meeting, we looked at the business of maintaining an optimal osmotic balance between the two major fluid compartments in insects, the intracellular compartment and the extracellular compartment.

We saw that the osmotic balance was, in part, achieved by mechanisms that restrict and offset the unavoidable loss of water that must necessarily influence terrestrial organisms. I emphasize the water loss is unavoidable. In particular, we saw that the properties of the cuticle, control of spiracular openings and resorption of water from faeces are important mechanisms to control water loss. There are also a variety of mechanisms to replenish water supplies.

Today we want to continue to develop our appreciation of osmoregulation by observing that organisms must an optimal balance of water and of salt within the body. We develop the problem by considering some specific data. A brief look at the concentrations of various ions in an insect's regular food and in its hemolymph shows a couple of points. First, the compositions of the the food and of the hemolymph are remarkably different in many cases. Second, and more important, the differences are not uniform across the range of individual ions.

Here are examples. Larvae of the bot fly *Gastrophilus* are internal parasites of horses, mules and some larger wild animals. In this insect sodium is much higher in the hemolymph than in the food, while potassium is higher in the food than in the hemolymph. A different situation obtains in the stick insect *Dixippus,* which often eats Privet leaves. Privet comes from shrubs in the olive family. Here, sodium, potassium and calcium are much higher in the food while magnesium is much lower. Finally, in the Colorado potato beetle *Leptinotarsa,* there is very little sodium in the common food while potassium and calcium are much higher in the food than in the hemolymph. This table presents the specific data:

Cation composition of hemolymph and food, given in milliequivelents/liter.

Insect species	Food	Sodium	Potassium	Calcium	Magnesium
Gastrophilus	Horse blood	17585	1231	62	323
Dixippus	Privet leaves	946	28152	16825	14240
Leptinotarsa	Potato leaves	4	65145	48129	18986

Now, if these salts were not somehow regulated, an insect would face the problem of osmotic, rather than water, imbalance between the two major fluid compartments every time it takes a meal. A clear example of this would be the large change in hemolymph calcium ions following a typical meal in the stick insect. So we want to ask: How are hemolymph ion concentrations regulated in the face of regular insult with each meal? Just imagine. One mechanism to regulate ion concentrations occurs at the level of the alimentary canal.

The insect would simply limit the absorption of unnecessary salts. They would absorb only the required organic food material while unnecessary inorganics were passed though the insect and voided in due course. However, that is not the way hemolymph ion concentrations are regulated. In point of fact, dietary inorganic salts are readily absorbed across the epithelium of the alimentary canal and the ions therefore mix with the general pool of inorganic material in the extracellular compartment. The physiological problem, then, becomes one of moving ions into and out of the extracellular compartment in such as way as to sustain an optimal osmotic balance.

The most general expression is that secretion occurs in the Malpighian tubules and resorption takes place in the rectum. A generalized model of the alimentary canal shows the foregut, midgut and hindgut. Within these familiar major regions are the Malpighian tubules, which most often join the alimentary tract in the midgut-hindgut junction, and the rectum, an enlarged portion of the hindgut featuring specialized epithelial cells. We want to appreciate the various sorts of movements of materials across the different regions of the alimentary tract. This is to be taken rather as a composite, and represents all sorts of movements, some of which do not occur in all insects. In due course we will see that water, inorganic ions, sugars and amino acids are mostly absorbed in the midgut.

With respect to ion concentrations, we want to see that water and ions are secreted into the lumen of the Malpighian tubules. These components are moved into the hindgut by mass flow, and then on to the rectum. Depending upon the individual needs of the insect, water and ions are reabsorbed by cells in the rectum and moved back into the hemolymph. We might get the impression that the same hemolymph constituents are moved in a sort of futile cycle, but we will see that the situation is otherwise.

Malpighian tubules are closed-ended tubules that arise from either the posterior part of the midgut or the anterior portion of the hindgut. The precise anatomical arrangement differs among insect groups. The walls of the tubules are one cell layer thick. The cells comprising the tubule walls stand on a basement membrane. The tubules from most insects are surrounded by muscles which move the tubules around in the hemocoel, and the tubules are supplied with numerous tracheae. Malpighian tubules of mosquitoes and some other insect species do not have muscles.

There are distinct histological and functional regions within the overall structures of Malpighian tubules from many, but not all, insects. The tubules have proximal and distal portions, proximal being closer to the attachment with the alimentary canal. The distal portion is functionally the secretory region, and features cells with honeycomb borders. By secretory we mean that water and ions are secreted from the hemolymph into the lumens of the tubules. The proximal region is composed of cells with brush borders; this is the absorptive region of Malpighian tubules.

We want to appreciate the cellular structure. The cells attach to the basement. Cells from the distal region feature a honeycomb border formed by cytoplasmic filaments. These filaments are rich in mitochondria, which supply the energy required for active transport processes. The cell membranes on the basal side are envaginated. All of these features are consistent with the secretory function of these distal cells. Cells from the proximal region of the tubule are different from the cells of the distal region in many insects. These cells feature a brush border. They are made up of more widely separated microvilli and have less complex envaginations of the basal-side membranes. They have more mitochondria in the basal side of the cell, thought to support their reabsorptive role in the tubule. Again, not all insects have Malpighian tubules of this heterogeneity. We also note that some insects, such as aphids, collembola and thrips have no Malpighian tubules or only small absorptive pads called papillae.

The Malpighian tubules of some insects have the distal ends adpressed to the sides of the rectum. The ends are surrounded by a membrane known as the perinephric membrane. The perinephric membrane forms a cavity between the rectum and the general hemolymph. This is known as a cryptonephretic arrangement. In the meal worm beetle *Tenebrio* this is thought to help absorb a maximum amount of water from feces.

A primary urine that is isosmotic to the hemolymph is formed in the Malpighian tubules. A general physiological observation is that water is rarely moved across membranes by itself. One way to remember this is to recall that ions can be moved by active pumping processes and water follows the locally formed gradient. In Malpighian tubules, fluid secretion is driven by active transport of potassium and chloride. In some blood feeders sodium is also actively transported. Potassium chloride (KCl) is usually the major solute in the lumenal fluid, and the concentrations of the ions are usually low compared to the hemolymph.

The Malpighian tubule fluid makes its way to the alimentary canal by bulk movement àided by muscles. The fluid then moves through the hindgut to the rectum. In the rectum selective reabsorption of water, ions and various metabolites occurs by active transport processes. It is only in the rectum that the urine becomes strongly different in osmotic concentration from the hemolymph. Both tubular secretion and rectal absorption are under separate endocrine controls which together determine the amount and ionic composition of water that is excreted.

The secretion and absorption processes are summarized in the figure. This drawing is derived from the blood-sucking bug Rhodnius. This represents those insects with heterogeneous Malpighian tubules. In the distal region of the tubules potassium, chloride and sodium are pumped into the lumen by active processes; water follows the ion gradient into the lumen. Note that the figure does not show the movement of chloride. We will talk about uric acid formation and excretion later, but for now please note that uric acid also moves into the tubule. Resorption occurs in the proximal portion of the tubule, with potassium moved by active processes, and water following. After movement of urine and uric acid crystals to the rectum, more water can be resorbed, along with selected ions.

Some insects have homogeneous Malpighian tubules, such as the stick insect also represented in the figure. Secretion takes place all along the homogenous Malpighian tubules. Some movements are active processes, such as sodium and potassium pumps, while others are passive. There is no resorption until urine is moved into the rectum, where several ions, including amino acids are moved back into the hemolymph by active processes.

Again, here are data to make the point. This information comes from work on the locust, all in meq/liter:

Osmotic pressure and concentrations of selected ions.

Compartment	Osmotic pressure	Sodium	Potassium	Chloride
Hemolymph	214	108	11	115
Primary urine	226	20	139	93
Rectal fluid	433	1	22	5

These data show specific reabsorption of ions on an individual basis. The hemolymph is typically high in sodium, while potassium is the major urine ion in the tubules. These data indicate that sodium and chloride are selectively resorbed in the rectum. The osmotic concentration of the hemolymph and the urine are about the same, although of differing ionic compositions. The fluid components of the rectal contents are much more concentrated than the other fluids.

These data come from locusts fed with fresh water. By way of helping to prepare for any future exams that could crop up right out of a clear blue sky, consider what these data might look like if the locust had been fed salt water. There would be high sodium levels in the rectum, as an avenue to excrete excess sodium. Again, ions are individually regulated and transported. The key point is that the Malpighian tubules secrete water and salts in urine formation. Superimposed upon this urine formation mechanism is a resorption mechanism to facilitate the selective uptake from the urine of those specific things required to sustain the optimal osmotic balance between the intracellular and extracellular fluid compartments.

Consider these situations that an insect may confront:

- Subjected to greater than normal desiccation: This situation may result in increased hemolymph osmotic pressure. The insect may respond with decreased urine formation, and greater rectal water uptake.

- Subjected to greater than normal desiccation and a food higher than normal in salt: This situation may result in increased hemolymph osmotic pressure, but the increased salt would not allow a maximal decrease in urine production. The insect may respond with a greater uptake of rectal water and a decrease in salt resorption.

From a biological point of view, Malpighian tubules in some insects are adaptive in unusual functions. Larvae of the face fly *Musca autumnalis* have four Malpighian tubules, two anterior and two posterior tubules. Both pairs arise from a single junction between the mid- and hind-guts. The posterior set is found in close association with the midgut. The anterior set expands distally, and is filled with a granulated liquid. The posterior Malpighian tubules are regarded as ordinary secretory organs.

The anterior tubules play a special role in the physiology of pupation. The puparium of this fly is hardened by deposition of calcium salts, rather than by the usual phenolic cross- linking typical of other insects. Dietary calcium salts, as well as other minerals, are stored in the anterior Malpighian tubules during the larval stages. At pupation, the calcium salts become dissolved, then transported from the tubules into the hemolymph. The calcium is directly transported from the hemolymph to the cuticle.

The Malpighian tubules of lacewing larvae undergo morphological changes before pupation. The modified tubules then produce silk for the pupal stage.

WATER BALANCE IN TERRESTRIAL INSECTS

We have spent a little bit of time considering aspects of movement. We've gotten an idea of the contractile process in muscle, a process driven by hydrolysis of ATP. We saw that ATP is generated as an end product of energy metabolism and considered gas exchange as a requisite for energy metabolism. Flight is very important to many aspects of insect biology, and I purposefully emphasize flight. Let us also emphasize the broader biological issue in flight. Flight can be taken under the mantle of motility, or movement. In this view, flight is one of many kinds of movement, all fundamental processes in biology.

Let us now turn attention to still another are of physiology confronting nearly every life form: osmotic balance. Osmotic regulation is the business of maintaining optimal salt and water concentrations in an organism. We begin by considering the biological significance of this. About 50 to 90 per cent of the body weight of an insect, or nearly any other animal, is water. Some of this fluid is inside the various cells that make up the body while the remainder is in the spaces outside of the cells. As a unit, these spaces are regarded as a collective known as the extracellular space. The extracellular space is filled with fluid, for obvious reasons called the extracellular fluid. Among the dissolved constituents of the extracellular fluids are the ions and the nutrients needed by the cells for maintenance of life. The extracellular fluid is in constant motion throughout the body, and it is rapidly mixed to near homogeneity by hemolymph circulation and by diffusion. Because of these mixing processes, we can say that all cells in the body live in essentially the same environment. The great French physiologist, Bernard, called the extracellular fluid the melieu int'erieur or what we would call the internal environment.

The point is to recognize two main fluid compartments in an insect body. One is the intracellular compartment and the other is the extracellular compartment. To appropriate a teaching device from Daniel Matzia, the famed cell biologist retired from UC Berkeley, we can regard the extracellular fluids as the *out environment*, that is, the environment inside the body that cells experience. The intracellular environment, then, could be the *in* environment. However, we know eucaryotic cells have organells and compartments within them.

Within cells, the compartment inside any particular organelle, such as a nucleus or mitochondrion, could be the *in-in environment*. At the organismal level, the world we live in is the *out-out environment*. This language allows us to appreciate the world in terms of environments that things experience: the in-in, the in, the out and the out-out environments. Cells experience the out environment, the same thing that Bernard called the *melieu int'erieur*.

Some percentage of the total body fluids of any organism are intracellular fluids. The fluid within each cell is made up of its own individual mixture of various constituents, such as ions, amino acids, proteins, sugars and so forth. On the other hand, within an organism, the concentrations of these things appear to be reasonably similar from cell to cell. For our discussion, the intracellular fluids of all of the different cells of an organism can be considered to be one large fluid compartment: the in environment or the intracellular compartment. Regardless of the particular language or expression that is easiest to keep in mind, we want to come to the point of view that an insect body has two major fluid compartments: the intracellular and extracellular compartments. The extracellular compartment is easy to visualize as the circulating hemolymph.

While the extracellular environment is substantially different in composition from the intracellular environment, the two compartments are in osmotic balance. That is to say that there is rarely a net flow of water from one compartment to another. There are various mechanisms to handle the maintenance of the correct concentrations of the intracellular constituents, including active transport and facilitated diffusion. These mechanisms are a related, but different matter. The thrust for our purposes is that water always tends to diffuse by a process called osmosis across cellular membranes from the side with fewer total dissolved particles to the side with a greater number of dissolved particles.

Let us consider some possibilities. First, if a cell in placed in a solution with far fewer osmotic particles than the cell, water will diffuse into the cell. This can result in considerable swelling, even to the point of bursting the cell. Alternatively, if a cell is placed into a solution with a far greater number of particles, water tends to diffuse, once again, down its osmotic gradient, in this case, out of the cell.

A badly shrunken cell that may or may not survive may result from this flow. We gather the significance of a physical phenomenon, osmosis. Osmosis explains the importance of maintaining a proper balance of water and of salts in the extracellular compartment or the melieu int'erieur. During times when, for whatever reason, the extracellular compartment is not correctly regulated, the physiological situation is exactly like placing cells into a solution that is too dilute or too concentrated. This is a common problem faced by insects and most other organisms living in freshwater, salt water or in terrestrial environments.

Again, terrestrial insects are faced with the problem of regulating their levels of body fluids and ionic concentrations such that the extracellular and intracellular fluid compartments are maintained in osmotic balance. Failure in this respect would lead to this situation: the body's cells are effectively placed in a fluid that would cause them to swell or shrink to a point of dysfunction or death. Physiological solutions to the problem of water balance have tremendous evolutionary and ecological significance.

First, it can be argued that the spectacular evolutionary radiation of insects occurred after a host of adaptations to terrestrial life. We saw some of these adaptations during our consideration of movement. Physiological regulation of water balance is another adaptation that supports terrestrial life. The biogeographical distribution of insects reflects the broad range of terrestrial habitats in which insects live and thrive. Many of these habitats are extremely arid and very hot. Physiological adaptations to these sorts of environments have permitted exploitation of such habitats that would otherwise be unavailable to insects. The focus on terrestrial life is not meant to overlook aquatic insects, which we will discuss in future.

Furthermore, aquatic insects arose from their terrestrial counterparts. Hence, adaptations to aquatic life can be thought of as a component of successful terrestrialization. Two main factors may work to upset the essential osmotic equilibrium within an insect's body:

1. Loss of water by transpiration across the cuticle, or a very large gain of water by drinking or eating.
2. The intake of food whose concentration of inorganic salts is quite different from the ionic composition of the hemolymph.

Today we are going to begin with the first point, the regulation of water balance, which may be insulted by substantial loss or gain of water. Insects are generally small creatures, which tends to put them in an unfavourable setting with respect to the ratio of body surface area to body volume.

The idea is to recognize a relatively large surface across which water can be lost by transpiration. We will see that water loss across the body surface can not be entirely stopped; however, if it were not restricted to a level that can be balanced by water uptake, insects would not survive in the more arid environments. The first line of defense was the development of a cuticle relatively impermeable to water.

The outer layer of an insect is the integument, which is made up of the epidermis and the cuticle. The epidermis is the outer cell layer. The epidermis lies on a basement membrante, and secretes the cuticle, which so far as I know is illustrated in virtually every book written on insect biology.

There are three major sections of cuticle, endocuticle, exocuticle and epicuticle. In standard drawings, we note the pore canal, cells of the epidermis and the basement membrane. The endocuticle provides the extensibility of the integument and combines toughness with flexibility. The rigidity of the hard parts of an insect, the appendages and so on, comes from the properties of the exocuticle. The relative impermeability of the cuticle to water is due to the presence of the epicuticle.

The epicuticle is a complex structure that differs among various insects. There is a cuticulin layer made of two components: a thin, dense outer layer and a thicker, still denser inner layer. Cuticulin is generally made of lipoprotein. A waxy layer lies atop the cuticulin.

This is made of long-chain hydrocarbons, and esters of fatty acids and alcohols. There is an inner wax layer composed of strongly oriented long-chain molecules that generally form a monolayer only one molecule thick. The strong orientation provides the waterproof properties of the epicuticle. The outer wax layer is composed of wax molecules that are randomly orientated; on top of this is the cement layer. People think that this is a shellac-like substance that may serve to protect the wax.

Terrestrial insects lose water by evaporation from the general body surface - a process called transpiration. The rate of water loss from the surface on any given insect is a function of both wind speed and also of the humidity of air. Transpiration is relatively independent of body temperature, up to some point. Water is lost from the body surface at a constant, but rather low, rate through a broad range of temperatures. At some temperature the cuticle reaches a temperature point called the transition point. This point varies from insect to insect, depending upon the chemical composition of the hydrocarbons and waxes in the epicuticle. Insects living in hot, arid habitats tend to have wax layers with higher transition temperatures than do insects living in cooler or more humid habitats. Some insects have two transitions points.

Water loss from the body surface is held to a low rate by the oriented wax layer on the outside of the cuticulin. When the temperature of the cuticle rises to such a point that the van der Waals forces holding the wax molecules together are broken, the molecules move slightly apart and water is allowed to escape. Below the transition temperatures the wax molecules are close together, and at a slight angle so that water loss is restricted. At temperatures above the the transition point, the molecules part, which usually leads to a quick and probably pleasing and comfortable death due to dehydration.

Then we see that osmotic balance in the melieu int'erieur is in part maintained by reduction of water loss from the body surface. Another important avenue of water loss in terrestrial insects is from the respiratory surfaces. In the orthopteran *Gastrimargus*, for example, 70 per cent of the normal water loss is from the trachael system.

We have already seen that water loss from the respiratory surfaces is restricted by the invagination, or internalization, of the trachael system such that it is inside the body. Water loss is also reduced by control of the spiracular valves. Internalization of the respiratory surfaces in a major adaptation to terrestrial life in vertebrates as well as in invertebrates.

We can model an experiment designed to demonstrate the influence of spiracles on water loss. Consider the rate of water loss from *Tenebrio* larvae as a function of time. *Tenebrio* is well adapted to arid conditions, and its water loss rates are typically quite low. If we impose the condition of open spiracles, by holding the animals in an atmosphere of 5 per cent carbon dioxide, water loss would be far higher than what could be easily balanced by an insect living in an arid habitat. Physiological regulation of the spiracles plays an important part in reducing potential water loss.

Some insects live through long periods of quiescence during which they do not feed or drink water. Control of the spiracles in these species is aimed at keeping them closed as much as possible to minimize water loss while facilitating the gas exchange necessary to support metabolic needs. One strategy is called cyclic release of carbon dioxide.

This has been well studied in pupae of *Hyalopohora cecropia*, a large insect suitable for study in this regard. Carbon dioxide is allowed to build up in the trachael system for a long as seven hours, then is released in bursts during which the spiracles are wide open. Between the bursts of carbon dioxide release the spiracles are not closed as tight as possible, but rather flutter so as to allow a continuous uptake of oxygen.

While the cuticle and control of the spiracles serve to restrict the loss of water, a certain amount of water is inevitably lost. Then we can identify three avenues of water loss:

1. Transpiration across the cuticular surfaces
2. Transpiration across respiratory surfaces
3. Loss of water in the excreta

Water that is lost must be balanced by an appropriate level of water uptake. Most insects take in sufficient water with the food they eat. In his book on insect structures and function, Chapman thought some insects may select food with a higher water content under certain circumstances. Still other insect species may consume more food than they actually need in order to realise the potential water in the food. Moreover, many insects drink water. The dipteran *Phormia,* for example, has specific water receptors in the tarsal and labellar sensilla. I've often observed wasps drinking water in desert pools in Mexico.

Some insects have evolved slightly more bizarre adaptations to gain water. Larvae of the dipteran, *Epistrophe,* can extrude an anal pipilla into a water droplet and absorb water. The cockroach *Periplaneta* is among many insects that can absorb water from water droplets on the cuticle.

The cuticle is asymmetrical with respect to water movement: water passes inwards more readily than outwards. Also, a few insects can absorb water from water vapour in the air when they are desiccated. *Tenebrio,* for example, can absorb water from air with relative humidities greater than 90 per cent. Prepupa of the flea *Xinopsylla* can absorb water from air at relative humidity above 50 per cent.

A final source of water is called metabolic water. As we saw during our look at metabolism, the final electron acceptor at the end of the electron transport chain is oxygen. Electrons and their hydrogens combine with oxygen at this point, yielding water.

The amount of water that can be gained by metabolism depends on the metabolic substrate. A hundred grams of the fatty acid palmitate (16:0) yields 112 grams of water while 100 grams of glycogen yields 56 grams of water. Grain insects, such as *Tribolium,* normally obtain a very large proportion if not all of their water from metabolism of their food. We can guess that metabolism of lipid may be adaptive with respect to water balance during long periods of hibernation in several insect species.

Chapter 7

Adaptations to Extreme Cold

Adaptations to environment, we can regard the material in several terms, including physiological ecology, enviornmental physiology or comparative physiology. As mentioned earlier, all these terms are meant to convey an interest in how insects are adapted to live in the habitats they occupy. Now let us move away from our interest in the importance and mechanisms of maintaining a homeostasis of water and salt balance. The adaptations to life in northern temperate and polar regions of our planet. We begin with the idea that insects inhabiting these regions are subject to freezing temperatures for a greater or lesser proportion of each year. Our main interest will lie in understanding how insects are adapted to very cold environmental temperatures.

One adaptation for life in freezing temperatures is to protect the body from freezing conditions. Many insects, and some fishes, protect themselves from freezing by biosynthesis of anti-freeze compounds. These compounds convey some degree of frost resistance, that is, an avoidance of freezing. The hemolymph concentrations of anti-freeze compounds increase and decrease in a seasonal pattern. The slug caterpiller *Monema flavescens,* is a good example. In this insect, concentrations of the polyol glycerol increase in hemolymph as winter approaches. As outdoor temperatures decrease in September there is a reduction in body sugar levels, which remain low throughout the winter. Attending this drop in body sugar is an increase in glycogen. Glycogen is nothing more than a long polymer of linked glucose molecules. When ambient temperatures drop to about 10°C, glycogen rapidly decreases in this species while glycerol increases to about 25 mg/g body weight. After winter, as environmental temperatures increase to 20°C, glycogen and glycerol concentrations are inverted again. In this example, increasing hemolymph glycerol to about 5 per cent of fresh body weight is associated with tolerance of freezing temperatures as low as -20°C. It appears that some insects can avoid freezing by generating anti-freeze.

The outcome of a well-designed experiment shows glycerol does, in fact, convey frost resistance to pupae of *M. flavescens*. Two groups of insects were treated. In one situation pupae were kept in chambers at 20°C. Insects maintained at this high temperature express no increase in hemolymph glycerol during the winter period. The low glycerol concentrations correspond to a relatively small increase in frost resistance, which reaches a peak in early November. The other group of pupae were reared outdoors through the fall. The increased glycerol concentration was accompanied by a dramatic increase in frost resistance.

Rates of increase in body glycerol concentration are often related to the ambient temperatures insects experience. We can consider the glycerol concentrations in several groups of overwintering *Papilio* pupae. Each group of pupae were maintained at a separate ambient temperatures. Pupae kept at 20°C experienced a small, baseline increase in body glycerol. Glycerol levels increase at higher rates when the pupae were kept at lower temperatures. Of course, glycerol is biosynthesized by enzyme actions, and at very low temperatures glycerol synthesis capability is reduced. For example, glycerol synthesis at 0 °C is faster than at -5 °C.

The next point we want to make is that glycerol is a common cellular metabolite, which we appreciate by regarding glycerol biosynthesis. The glycolytic pathways were presented in the treatment of energy metabolism. The aldolase reaction is a key step in this pathway. Aldolase splits one 6-carbon compound, fructose-1,6-diphosphate, into two 3-carbon compounds, dihydroxyacetone phosphate (DHAP) and glyceraldehyde-3-phosphate. The step from DHAP to glycerol-3-phosphate (G-3-P) is very common because this compound provides the backbone structure for biosynthesis of triacylglycerol, with which we are comfortably familiar. The phosphate is removed from G-3-P by phosphatase, yielding glycerol. My speculation is that glycerol synthesis is regulated at the phosphatase step. Overall, glycerol is just two metabolic steps away from DHAP, a common intermediate in the glycolytic pathway.

In spring months glycerol levels are reduced, with a concomitant increase in glycogen. This is arranged by simply reversing the steps just covered in glycerol biosynthesis. The aldolase reaction is commonly reversible, as are most of the steps in glycolysis. The flow of carbon from glycogen → glucose → glycerol and back again glycerol → glucose → glycogen is a straight forward metabolic pathway. The key point is recognizing the carbon atoms are not lost, they are simply moved around some.

It seems appropriate to ask: How does glycerol increase the cold hardiness of an insect? The answer is glycerol probably works in three ways. One is freezing point depression, due simply to colligative properties of molecules. A second is supercooling. Finally, glycerol provides protection from ice damage to tissues and cells. First, we recollect from basic chemistry courses what are known as colligative properties of solutes. By this we mean those effects of a solute that express their influence by the number of particles, rather than by the nature of particles. For example, a 1.0 molal solution (1 mole of solute in 1 liter of water) lowers the freezing point of water by 1.86°C; it also raises the boiling point and creates other physical sorts of events.

The influence of all colligative properties can be calculated from the influence of any other colligative property. The idea for our purposes is that one effect of increasing hemolymph glycerol concentrations is to lower the freezing point of body fluids due to the colligative properties of solutes. Putting this together with our information on regulation of osmotic balance, it should be clear that colligative properties alone would not be a very effective overall strategy to avoid freezing.

The second mechanism allows lowering temperatures of body fluids below freezing points without formation of ice crystals. This is called supercooling. Let us consider a single example. First, animal body fluids are not pure water, and animal freezing points are typically a bit lower than 0°C. Second, with supercooling, body fluids can decline considerably below zero without freezing. Finally, a supercooling point obtains. This is the point at which supercooled fluids freeze. We usually see a small temperature rise at this point, due to heat of fusion.

We can consider a graph illustrating supercooling in pupae of *Papilo xuthus*. When these pupae are cold hardened they can be cooled to about 20°C below the usual freezing point without ice crystal formation. The temperature at which the fluids finally do freeze is called the supercooling point. This event is associated with a sudden rise in body temperature due to heat of fusion. On a larger scale, the temperature increase due to heat of fusion is why very large lakes exert a moderating effect on environmental temperatures in some climates. Many insects from northern latitudes can be supercooled to 20 to 30°C below 0°C. In a more extreme case, larvae of the braconid wasp *Bracon cephi* can be supercooled to -47°C.

Some insects appear to actually freeze because environmental extremes far exceed supercooling capacities. In these cases glycerol also seems to protect tissues and cells from injuries that otherwise obtain due to ice crystal formation. This non-specific protection from ice crystal formation is well known. Glycerol has been used for decades in laboratories to protect vertebrate tissues during freezing. For example, human sperm can be kept frozen, and viable, for years if glycerol is added before freezing. Freezing is lethal to sperm that are not protected with glycerol.

Glycerol exerts three actions in cold hardiness. Glycerol acts as a cryoprotectant by lowering freezing points of body fluids, by enhancing supercooling capacities and by protection from ice-crystal formation. Glycerol is not the only substance seen in studies of cold hardiness. The other most commonly found compound is the 6-carbon polyol sorbitol. Sorbitol is formed by a single step reduction of glucose-6-phosphate, followed by removal of the phosphate group. The sortibol mode of action within a biological matrix appears to be much like the action of glycerol. Beyond glycerol and sorbitol, a few insects increase trehalose concentrations for onset of winter. Still other species biosynthesize mannitol, another polyol.

Although it is certain that glycerol can exert an important cryoprotectant influence on many insects, there still exists a certain ambiguity as to its precise role. Looking at some data on the point, we find examples of three cases: In one case, for example the lepidopteran mentioned earlier, Monema flavescens, the prepupae accumulate glycerol (to about 5 per cent of fresh weight) and gain protection from freezing. In a second case, insects accumulate glycerol, but they do not gain protection. This occurs in extreme cases in some species. For example, larvae of the beetle *Dendroctanus manticolae* accumulate up to 23 per cent of fresh body weight in glycerol and gain no protection from freezing. In a third case, insects do not accumulate glycerol, but they are able to tolerate very cold ambient temperatures. Larvae of the lepidopteran *Hestina japonica*, for example, can tolerate freezing to -15^{oC} with no increases in body glycerol.

These data are not presented to suggest that the formation of glycerol does not play the frost resistance roles we just considered. Rather, we want to take the idea that other, yet unspecified, biochemical activities appear to act in the physiology of cold hardiness.

Now, the physiological mechanisms we have looked at so far suggest that insects either build up anti-freeze compounds, or they develop a capacity to supercool, which may or may not involve glycerol. These strategies are known as freezing resistance. Another strategy is known as freezing tolerance. Freezing tolerance is the ability to survive formation of ice within an organism. Freezing tolerance is known in a number of insect species. The larvae of one species of arctic midge, for example, overwinters in solidly frozen pools. These larvae begin to freeze at –1 to –2 °C. They can be frozen to -25oC and subsequently thawed without injury to the tissues. They can, in fact, tolerate many cycles of freezing and thawing. In the neighbourhood of –15 °C, about 90 per cent of this midge's body water is frozen. How can some insects tolerate freezing?

One explanation is called the site of freezing theory. The first point is that intracellular freezing is fatal, whereas extracellular freezing, to some extent, is not necessarily fatal. A second point is that freezing usually begins extracellularly, with any freezing within intracellular compartments occurring subsequent to extracellular freezing. This is because the formation of ice crystals depends entirely upon the chance of an aggregation of water molecules getting to the critical size of a crystal nucleus; the probability of such nucleation increases with the bulk volume of water, which is greater in extracellular compartments.

As ice develops around the cells, an increase in concentration of extracellular osmotic particles attends the decrease in free water. This is followed by movement of water from the intracellular compartment into extracellular compartment, thereby increasing solute concentrations within cells. Loss of intracellular fluid has two effects. First, the increased solute concentrations makes cells more resistant to freezing. Second, if a cell does freeze, the resulting ice crystals are much smaller, and they do less physical damage to intracellular organelles.

One prediction that is consistent with the site of freezing theory is that the rate of cooling could effect outcomes of freezing experiences. Here are some data from the sawfly *Trichiocampus*. In the following experiments all insects were cooled to –20°C, but at different rates of cooling.

Rate of Cooling	Outcomes
327°C/min (immerse in liquid N_2)	All fat body cells frozen; all larvae dead upon thawing
4°C/minute	Most, but not all cells frozen; thawed larvae looked normal, but did not pupate
0.8°C/minute	17% of cells frozen; 50% of thawed larvae pupated and produced normal adults
0.4°C/minute	No intracellular freezing; all thawed larvae produced normal adults

These data are consistent with the idea that insect physiologists do some pretty weird experiments. The data also show that when cooled very rapidly, the intracellular and extracellular body fluid compartments can freeze at about the same time, without a crucial time lag between initial freezing in extracellular compartment and later freezing within cells. The time lag would allow for osmotic movement of water from intracellular to extracellular spaces. It is thought the deleterious effects of intracellular freezing in the fully hydrated state result from mechanical damage. Expansion of ice crystals within cells would disrupt cellular organization, injure organelles and possibly rupture membranes. Slower cooling rates allow time for water to move out of cells before they freeze, thereby lessening potential for physical damage.

We can infer a little bit more from this information. At least part of the ability to survive freezing is related to ability to withstand dehydration. The most extreme example of the ability to survive at very low temperatures comes from studies on larvae of the midge *Polypedilum vanderplankii*. These larvae are active in temporary pools in tropical Africa, an area little supposed to have much to do with freezing. The temporary pools are regularly subject to complete dehydration, and the larvae are adapted to survive loss of up to 92 per cent of their body water. In studies incidental to this primary adaptation, it was shown these larvae can be cooled to -270°C and immediately resume activity upon being thawed and rehydrated.

Most insects living in high mountains or in cold temperate regions are well adapted to resist or to tolerate freezing. Nonetheless, the most extensive adaptations to cold environments are seen in the relatively few insects inhabiting the high arctic and a few islands in the southern polar regions. In these areas of relatively harsh circumstances, with low temperature, high winds and brief periods in which to affect mating and growth, freezing tolerance is but one of several adaptations to special situations.

The most extensive specializations appear to occur among members of the Simuliidae, commonly known as the black flies. In ordinary, off-the-shelf southern species, females fly in search of blood meals from avian or mammalian hosts. After feeding females settle down until blood is digested and eggs are matured. Eventually the females embark on oviposition flights, following which another cycle of feeding, egg maturation and oviposition may begin afresh. A female may complete several such cycles during her adult life. Flight is also important in mating, because males and females meet in swarms where males respond to females that are detected visually through enlarged facets in upper parts of their compound eyes.

Unlike their less extreme relatives, most the adult female of most arctic species is a modified non-feeding form. Mouthparts in these species are so reduced that it is not possible to penetrate host integuments. Their eggs are developed on nutrients that were accumulated in larval stages. In fact, eggs start to develop before adults emerge from pupal cases. In some species the eggs are mature at adult eclosion. These changes in developmental patterns have the effect of telescoping the life cycles. Moreover, mating occurs as a result of contacts made on the ground near sites of emergence, rather than in a swarm. Mating chances are enhanced by minimal dispersal of adults and by dense larval aggregation in immature stages. Often the enlarged eye facets of males are reduced. The main points are reduction of feeding and mating flights, early egg development and re-designed mate-finding systems. These adaptations reach even greater extremes in a few species: populations occur as male-less pathenogenic reproductive females.

Other dipterans inhabiting the tundra regions express similar adaptations. The biting midge *Culicoides* is autogenous, that is, it does not require a blood meal for egg maturation, and their mouthparts are slightly reduced. Also, in most arctic Chironomidae mating takes place on the ground without a mating flight. There are also examples of parthenogenic races. In arctic Tipulidae the wings are reduced to a greater or lesser extent; females of one species have only vestiges of wings.

Arctic mosquitoes tend to retain their normal appearances, but most species are facultatively autogenous. Two species of *Culiseta* hibernate as mated, but not blood-fed adults while some *Aedes* species hibernate as eggs. Even though most of these species can mature their eggs in the absence of a blood meal, there is reduced fecundity associated with this trait.

The most striking feature of arctic lepidopterans is the long length of larval lives. Larvae may develop over a period of five years or more before finally pupating. These larval stages are remarkably tolerant of extreme cold and exposed habitats in winter. Other insect groups are represented in arctic regions, including aphids, scales insects, lice, bumblebees, flightless beetles, mayflies and caddisflies. A few common ideas run through the adaptive complex of all insects found in arctic regions. Flight in unstable thermal conditions that are punctuated by high winds is a very hazardous business. Many of the adaptations noted here are well suited for reducing the importance of flight. Host and mate finding may be very difficult under these extreme conditions, and certain points tend to minimize these requirements.

A final point is that these insects, as well as many other temperate species, share the ability to survive in temperatures well below freezing.

AQUATIC ENVIRONMENTS

One perspective on the physiology of organisms is to regard the question of how organisms are physiologically adapted to live and thrive in the particular place they happen to occupy. Elimination of ammonia in the form of uric acid, for example, is regarded as one of several physiological or biochemical mechanisms that help insects, as well as reptileS and birds, make their living in very arid environments. We will address nitrogen metabolism later in the series. The ability to maintain thoracic temperatures well above ambient temperatures is another physiological adaptation that enables many insects to operate in thermal regimes that would otherwise restrict activity. Physiological ecology directs attention to those areas of physiology that can be perceived as a sort of template, variations upon which can convey special adaptations to environment. These areas can be seen in contradistinction to other areas of physiology, ones relatively fixed across a wide range of phylogeny and environment. The physiology of protein targeting is a good example of a process seen in most eukaryotic cells. The study of protein targeting would not fall under the mantle of physiological ecology.

Aquatic insects share with their terrestrial counterparts the problem of maintaining an optimal osmotic balance with respect to their intracellular and extracellular body fluid compartments. Recalling a figure from an earlier discussion, we review the potential effects of osmotic imbalance. Cells placed in an environment of far higher salt concentration than themselves often lose water, becoming shrunken and deformed. To the other side, cells placed in an environment of far lower salt concentration than themselves often gain water, even to the point of bursting. Again, maintaining an osmotic balance is important to survival of organisms.

The situation for aquatic insects is easy to apprehend. Insects living in fresh water are generally hyperosmotic to their environment. That is, extracellular compartments of aquatic insects have higher osmolyte (ions and amino acids) concentrations than the surrounding water. Under this condition water tends to diffuse into aquatic insects. If left unattended this process would effectively place cells in a very dilute environment. Insects living in salt water, such as brine flies, are generally hypoosmotic to their environment. These insects tend to lose water to the environment. Again, if left unattended, this would put the cells within the insect in a state of water loss and shrinkage.

Equipped with a feeling for the nature of the problem, let us begin with a consideration of fresh water insects. The hemolymph, or extracellular compartment, of these insects is hyperosmotic to the water comprising their habitats. Water tends to pass through the cuticle and into insects. Some adult forms, such as the predatory beetle *Dytiscus*, have cuticles relatively impermeable to water. However, the majority of aquatic insects in their larval stages are quite permeable to water. Often respiratory gill structures are even more permeable to water, and additional water is taken in with food. We gather that fresh water insects have a problem of taking in too much water. The relatively large amounts of water invading fresh water insects are removed from their bodies by formation of copious amounts of dilute urine. What we want to know, then, is how the urine is produced.

The timing of this lecture is designed to provide a review of the general information seen just lask week, while adding a bit more information on osmoregulation generally. We begin by revisiting the structural features of the insect renal system. We can take a look at the figures from the previous lecture. This figure shows Malpighian tubules arising from the alimentary canal. Again, these are thin, close-ended tubes floating free in the hemocoel where they are continuously bathed in hemolymph. A single layer of cells is attached to a basement membrane; often muscles spiral about the outside of this membrane, and the tubules can wiggle within the hemocoel. Small tracheae run along the length of the tubes, which supply oxygen to mitochondria located in muscles and in Malpighian tubules. Fluid within the lumen Malpighian tubules is isomotic to the hemolymph, but of different composition. Hemolymph is generally high in sodium, chloride and amino acids, while the intratubular fluid is high in potassium.

The general expression is that potassium is actively pumped from hemolymph to tubule lumens and water passively follows the ion movement. Recall there are homogeneous and heterogeneous Malpighian tubules. They differ in resorptive properties, but the general patterns apply to both types of Malpighian tubules. The flow of water into Malpighian tubules increases local hydrostatic pressure within the tubules, resulting in a bulk flow of fluid. This is the primary urine flow. The primary urine is moved into hindgut, then to rectum. After selective resorptive processes in the rectum, the fluid is voided into the fresh water environment.

Then we can imagine water moving from outside insects into their bodies by crossing their cuticles, and in some inseccts, the gills. The water is transported to rectums by way of Malpighian tubules at considerable expense of metabolic energy. The energy cost is related to actively pumping potassium ions. The active pumping processes also help us understand the many tracheal connections to Malpighian tubules.

Let us allow ourselves a slight diversion for just a minute. We want to be sure our language is clear. When we consider any two fluid compartments, three and only three osmotic conditions can exist. One compartment can be more dilute than the other, or hypo-osmotic. A compartment can be exactly the same as the other compartment, or isosmotic. Finally, one compartment can be more concentrated than the other compartment, that is, hyperosmotic. You will recall from your mathematics courses that any number can only be less than, equal to, or greater than another number. There are no other possibilities. Then what we need to keep in mind is that an osmotic situation of one compartment is always relative to another. Hence, fresh water is likely to be hypoosmotic to an organism and the organism would be hyperosmotic to the fresh water.

We are considering insects living in fresh water. In this situation ions inside insects tend to move down their concentration gradients to their relatively dilute, or hypoosmotic environment. There is a general requirement that this loss of ions be offset. More immediate to our story, however, is the observation that if the potassium ions in the urine were not somehow recovered, an insect would soon be unable to continue excreting excess water. Why would this be so? Because the movement of excess body water follows passively from the active movement of potassium ions. The physiological solution, as you already well know, is that the complete excretory system of insects includes the rectum. Secretion is by way of Malpighian tubules and reabsorption by way of rectum. Here are some data from larvae of the trichopteran *Limnophilus* reared in a salinity equivalent to 0.01 per cent NaCl.

The osmolarity of the extracellular compartment and of the fluid within the Malpighian tubules is about 0.031 per cent NaCl equivalents. The osmolarity of the rectal fluids is about 0.009 per cent NaCl equivalents. Similarly, in the fresh water mosquito larvae of *Aedes aegypti* we find something like 0.63 per cent NaCl equivalents in the extracellular compartment and 0.57 per cent NaCl equivalents in the fluid within the Malpighian tubules. Again, the rectal fluid is considerably different at 0.07 per cent NaCl equivalents. In both of these exemplars the rectal fluids are very dilute relative to the Malpighian tubules and hemolymph.

What can we make of this information? The general idea is potassium is actively pumped into the lumen of Malpighian tubules. Water accompanies this movement, along with some amounts of other salts, and there is a concomitant bulk flow into hindgut and thence to rectum. Before the water is voided, salts are again actively pumped back into the extracellular compartment by specialized regions of rectum. This second pumping of ions produces a very dilute urine which can be voided while ions are kept within bodies.

Salts are lost across cuticles and, to lesser extent, in urine. The physiological consequence of losing body salts is similar to gaining excess water: extracellular compartments become too hypo-osmotic with respect to intracellular compartments, thereby presenting a deleterious osmotic imbalance between the two. Insects living in fresh water, then, actually have two problems. First, they must produce copious amounts of dilute urine so that extracellular fluid compartments are not diluted by incoming water. Second, they must somehow replace salts that are inevitably lost, also to keep extracellular compartments from becoming too dilute.

Some salts will be gained from food, but in addition to this, some aquatic larvae are able to take up salts from very dilute solutions in the water they live in. Larvae of the dipteran genera *Aedes, Culex* and *Chironomus* take up salts though specialized structures called anal papillae, also known as anal gills. This uptake requires energy expenditure. The enery is exerted in pumping sodium, potassium, chloride and phosphate ions against gradients from dilute fresh water to hyperosmotic extracellular compartments.

The anal papillae hypertrophy as media become more dilute, presumably giving greater surface area for transport of salts. Some insects are able to extract ions from very dilute solutions. For example, larvae of the yellow fever mosquito *Aedes aegypti* are able to maintain steady osmotic states in media containing only 1 X 10-6 moles per liter of sodium.

Other aquatic larvae, such as the neuropteran *Sialis,* have no specialized surface for salt uptake. These larvae produce a urine very low in salts, but higher in other osmolytes, mostly ammonium carbonate, as shown here:

Ionic Composition

Fluid	Na	K	Cl	NH_4	HCO_3
Hemolymph	10	5	31	- -	- -
Urine	12	4	0	106	91

In addition to their ability to resorb salts in rectums, these insects are able to compensate for osmotic imbalance by maintaining higher levels of hemolymph amino acids. These amino acids serve as osmolytes and thereby help maintain the crucial balance between the major fluid compartments. This emphasizes the point that osmotic compositions of the two major fluid compartments can be quite different while osmotic concentrations can be quite similar.

In contrast to fresh water insects, those living in salt water face problems more like terrestrial insects. Their hemolymph is often hypoosmotic to their media and they tend to lose body water across their cuticles. We know that this can lead to extracellular fluid compartments that are hyperosmotic with respect to intracellular compartments. Water lost to hyperosmotic environments can be replaced by controlled drinking of saline media followed by physiological uptake of water across midgut epithelia and from rectums, yielding a very concentrated urine. Salt water ingestion can disrupt midgut cells in fresh water insects, but salt water insects can drink saline water without injury. Here are some data to clarify the point:

	Equivalent % NaCl		
	Urine	**Hemolymph**	**M.t.**
A. aegypti (dH_2O)	0.07	0.63	0.57
A. detritus (sea H_2O)	3.50	0.93	1.20

These are data from larval mosquitoes. Although both species are in the same genus, *Aedes,* one species lives in fresh water and the other lives in brackish water. What we might see here is that the hemolymph in both species is fairly similar in osmotic concentration, while the urine of the salt water species is much more concentrated than the urine of the fresh water species.

It is appropriate to summaries the basic physiological adaptations to aquatic life:

Parameter habitat	**Aquatic habitat type Freshwater habitat**	**Saltwater**
Ionic concentration	Hypoosmotic	Hyperosmotic
Extracellular compartment	Hyperosmotic to environment	Hypoosmotic to environment

Water movement	From media to body	From body to media
Ion movement	Lose body salts	Gain body salts
Malpighian tubule fluid	Isosmotic to hemolymph	Isomotic to hemolymph
Rectal fluid	Hypoosmotic to hemolymph	Hyperosmotic to hemolymph
Rectal uptake	Salts	Water

Let us look at the nature of aquatic environments just a little more closely. Fresh water habitats, depending upon the local geology, may have total ion concentrations ranging from less than 0.1 mM to more than 10 mM in NaCl equivalents. Moreover, salt concentrations can vary widely due to evaporation. In addition to total salinity, the relative amounts of particular ions may vary over a considerable range. This is markedly true in waters with unusual substrates such as limestone. Likewise, salt water habitats may be extremely variable in ionic composition. Brackish water occurs in coastal regions where sea water and fresh water are mixed. Tidal flows can vary salinities of such areas over a wide range of concentrations. Because environments are variable, we achieve far greater insight into physiological adaptations of aquatic insects if we know the range of salinity that insects are adapted to tolerate.

Organisms that can tolerate wide variations in salt concentrations in the water they live in are called euryhaline from the Greek words for wide and salt. An extreme example is the brine fly *Ephydra* which lives in the Great Salt Lake. These flies can tolerate periods in distilled water and periods in water concentrated to 20 per cent NaCl equivalents. In low environmental salt concentrations the larvae are able to hyperosmoregulate, that is, able to maintain extracellular fluid compartments at higher osmotic concentrations than the medium. When their environment becomes highly concentrated these larvae can hypoosmoregulate, or maintain extracellular fluid compartments at an osmotic concentration lower than their environment. These insects are adapted to tolerate a very wide range of salt concentrations in their environments.

At the other end of the spectrum, some aquatic insects are stenohaline, indicating a narrow range of tolerance to variations in environmental salinity. Larvae of the mosquito *Aedes aegypti* represent a good example. These larvae maintain hyperosmotic hemolymph concentrations over a narrow range of salt concentrations. When environmental salt concentrations exceed osmotic concentrations in hemolymph, however, control breaks down, and the larvae become osmoconformer. Their hemolymph salt concentrations go far beyond appropriate levels, and the larvae fail to survive.

Many insects are adapted somewhere between these extremes of highly euryhaline and very stenohaline. These insects are able to hyperosmoregulate in fresh water and to a limited extend they can hypoosmoregulate in slightly brackish water. Beyond some limit of concentration, control is lost and the larvae become osmoconformers. You can appreciate the adaptive significance of nonconformity.

PHYSIOLOGY OF DIAPAUSE

During our last meeting we considered one aspect of living in temperate and polar regions of our planet, that is the business of cold hardiness in insects. We emphasized mechanisms of resisting and tolerating freezing. Biosynthesis of anti-freeze compounds allowed body fluids to cool far below a normal freezing point without ice crystal formation. This process is known as supercooling. In addition to supercooling, certain insects can tolerate freezing, for which the most important adaptation is that they can tolerate a remarkable level of cellular dehydration. These physiological adaptations to life in very cold conditions can be taken as rather extreme examples of a sort of biological recognition that conditions in almost every environment, aquatic, marine, terrestrial or atmospheric, are subject to regular and irregular changes. Today we want to regard some of the kinds of changes in environment, and look at some of the ways insects respond to changes.

Many insect species adapt to regular, seasonal changes in environmental conditions by manifesting various expressions of diapause. We want to appreciate what diapause is, and touch upon some aspects of physiological regulation of diapause. Tauber developed a comprehensive definition of diapause:

Diapause is a neurohomonally mediated, dynamic state of low metabolic activity. Associated with this are reduced morphogenesis, increased resistance to environmental extremes, and altered or reduced behavioural activity. Diapause occurs during a genetically determined stage of metamorphosis, and its full expressions develops in a species-specific manner, usually in response to a number of environmental stimuli that precede unfavourable conditions. Once diapause has begun, metabolic activity is suppressed even if conditions favourable for development prevail.

This is a large definition, and it is meant to convey several points: First, diapause is not a single physiological event, but rather is a syndrome of several physiological and behavioural manifestations all bearing on adapting insects to forthcoming environmental changes.

Second, the environmental signals that are translated into a diapause syndrome are called token stimuli because they do not themselves impact on insect life. Token stimuli simply herald the coming changes.

Third, the definition is biologically large because it includes dormancy and seasonal migration and seasonal polyphenism. To keep our terms straight, let us agree that dormancy is a period in a life cycle during which growth, maturation and reproduction are held in check. Dormancy can occur in any season. Polyphenism refers to seasonal changes in colour or structure of an organism. There are many examples of this among insects, including increased wax secretion, variations in wing length and quality of cocoons (tighly versus loosely spun).

Fourth, the events in a dispause syndrome are not aimed directly at organismal survival during any particular set of environmental conditions. Diapause has more to do with synchronization of entire life cycles to seasonal changes in environments. It affects survival during unfavourable conditions, but it also bears a large weight upon other features of life history, including growth, development and reproduction. Hence, diapause is a unifying biological concept. Because it is under genetic control, it is an important substrate in evolution of life histories and in speciation.

Insects also cope with unpredictable environmental changes, such as droughts and overpopulation. Short-term or localized changes might include sudden temperature changes or local food shortages. Insects can adjust to these sorts of changes with aseasonal quiescence or aseasonal migration. These adaptations differ from diapause in the manner of physiological control. We will see that diapause is controlled by endocrine regulation; aseasonal adaptations are under central nervous system control.

Diapause is a genetical feature of insect life histories. All metamorphic stages in life cycles can be utilized as a diapause stage, but diapause usually occurs in a stage characteristic of a given species. The definition that diapause is a phenomenon that anticipates future seasonal changes, we see that token stimuli must be perceived by an insect, then translated into a biological response. These token stimuli are perceived only during specific stages, and as with the diapause itself, timing of perception of token stimuli is under genetic control.

Perception of token stimuli and expression of diapause can be rather close in time, or widely separated between two generations. Here are some examples of the possibilities. Many insect species undergo diapause during the egg stage. In the cricket *Teleogryllus,* young eggs are sensitive to diapause-inducing stimuli, and diapause follows in eggs that are slightly further along in development. There are also cases in (for example, a cricket and two lepidopteran species) in which diapause-inducing stimuli are perceived during various stages in development of the mothers. Finally, in the aphid *Megoura* the grandparental generation is the sensitive one.

Let us consider diapause in context of overall seasonal cycles. First, diapause has recognizable stages. The *prediapause period* is regarded as a time of feeding and metabolic reserve accumulation, as well as growth to the species-specific diapause stage. The *induction period* is usually characterized by reduction in feeding, metabolism and growth. The *maintenance period* sets the length of diapause, and is thought to be genetically regulated. During this time insects are sensitive to token stimuli, which can affect the intensity and length of diapause. These two remarks are not to be taken as contradictory, but rather as an indication of a complex regulatory system. Diapause is broken by token stimuli, which entrains a series of physiological events leading to increased metabolism and post-diapause transition to non-diapausing condition.

Going back to the idea of a diapause syndrome, we want to emphasize that diapause is not simply a reduction in metabolism and growth, but is a combined expression of several related, but different, physiological adaptations. These can include prediapause behaviours (such as preparing a hibernaculum), accumulation of glycerol for cold-hardiness, and colour changes to track changes in plant colours, as well as alterations in energy metabolism. Insect endocrine systems play a central role in all of these events. At one time it was thought that diapause was mediated by a single endocrine mechanism for each life stage, but, as usual, the situation is more complicated.

It is generally agreed that diapause in adult insects attends reduced corpora allata activity, and hence, reduced hemolymph concentrations of juvenile hormone (JH). As the name suggests, we usually regard JH in terms of insect developmental physiology. However, its actions in regulating developmental events in juvenile insects is just one aspect of JH biology. JH is normally present in adult insects. JH is involved in regulating aspects of sexual maturation, such as egg development in females of some species and accessory gland development in males of some species. It follows that reduced JH titres (in endocrinology, titre refers to circulating levels of a hormone) would at least impact on reproductive aspects of diapause. In some cases adult diapause can be terminated by treating diapausing adults with JH.

Adult diapause is induced by short day lengths in the Colorado potato beetle. Diapause can be induced in beetles maintained under long day conditions by extirpating the corpora allata. In this species corpora allata activity is regulated by stimulatory neurohormones from the brain. Termination of diapause, accordingly, requires long day conditions. The sources of the stimulatory neurohormones are elements of the pars intercerebralis and corpora allata. Hence, in the Colorado potato beetle complete regulation of diapause involves neurohormones and JH. Just to give you a more detailed picture, the case is more complicated still in this beetle. Individual features of the diapause syndrome appear to be regulated by specific titres of JH, so that JH probably plays several different regulatory roles.

Diapause is under slightly different regulatory mechanisms in the bug *Pyrrhocoris*. Here JH and neurohormones are involved, as in the potato beetle. This bug differs, though, because there are inhibitory neurons from the brain that directly inhibit corpora allata activity, and thereby maintain low JH titre and diapause. The point here is recognizing an element of direct neurological control of diapause in this insect.

The classic work on pupal diapause was carried out on the giant silkworm, *Hyalophora cecropia*. When given their choice, physiologists often work on larger insects. The work on the giant silkworm was carried out by Carrol Williams and his co-workers at Harvard University. I guess you might say at the Nebraska University of the east coast. In this species pupal diapause follows cessation of brain neurosecretory activity, which stops prothoracicotropic hormone (PTTH) release. When we treat the physiology of post-embryonic development we will see PTTH stimulates prothoracic glands to release ecdysteroid, the moulting hormone. For our current interest in diapause, we note after a period in cold temperatures, "reactivated" brains release PTTH. PTTH stimulates release of moulting hormone, and this hormone signal leads to active adults.

The pupal diapause of another lepidopteran, *Heliothis zea*, is independent of PTTH. Low temperatures somehow act directly upon prothoracic glands, which reduces titres of moulting hormone. Pupae of flesh flies break diapause in response to moulting hormone, but the precise mechanism is slightly different from the lepidopterans just discussed. In flesh flies a sharp but brief peak of JH occurs before release of ecdysteroid. The JH is thought to positively influence, or somehow prepare, tissues to respond to ecdysteroid.

Larvae of many insects undergo diapause. In some speces, induction and maintenance of diapause depend on continuous presence of JH. In the rice stem borer, *Chilo suppressalis,* JH titres remain high throughout diapause, and corpora allata activity decreases as diapause draws to an end. This physiological mechanism of regulating diapause occurs in other species, too, but does not explain diapause in all insect larvae. In another mechanism, some diapausing larvae continually secrete JH from active corpora allata. In these species, diapause is induced and maintained because the JH inhibits release of PTTH.

Diapause interupts embryological development at a very early stage in the eggs of some insect species, such as the silkmoth *Bombyx mori*. This results from a maternal diapause hormone produced by the mother. Maternal diapause hormone is produced in the subesophageal ganglia, then transported via hemolymph circulation to eggs within the ovary. Circulating maternal diapause hormone causes diapause in eggs. A similar mechanism may occur in other insects, including the bug *Adelphocoris*.

Maternal diapause hormone is released from the subesophageal ganglia. The hormone release is triggered by enviornmental events. Silkworm females which experienced long days and warm temperatures during their egg and early larval periods lay eggs that enter diapause. Females which experienced short days do not produce maternal diapause hormone, and their eggs do not enter diapause. The maternal diapause hormone of *Bombyx* has been purified. It is a dipeptide, consisting of two different peptide chains.

Diapause hormone acts to arrest embryonic development at the stage of germ band formation. There is a tight correlation between diapause and carbohydrte metabolism. The developing ovaries accumulate large amounts of glycogen during the induction phase of diapause. When diapause begins, the glycogen is converted to polyols. At the end of diapause the polyols are converted back into glycogen, which becomes the major source of energy to carry out embryonic development. Diapause hormone is thought to influence carbohydrate metabolism in eggs.

The follicle cells of the ovaries are the targets of diapause hormone. Let us consider a model of diapause hormone action. Diapause hormone is thought to interact with specific receptors on the plasma-membrane of follicle cells. The hormone receptor translates the hormone signal into reduced intracellular levels of cGMP. The decreased cGMP levels activates increased membrane trehalase activity. This activity leads to increased trehalose hydrolysis and in-bound transport of glucose. The glucose is rapidly taken up into glycogen.

Beside its effect on carbohydrate metabolism, diapause hormone is thought to exert other actions on ovaries. The physiology, ecology and evolution of insect diapause remain contemporary areas of study. A far more complex picture of hormonal regulation of diapause will emerge in due course.

Chapter 8

Cockroaches

Cockroaches are flattened, fast-running, nocturnal insects that seek warm, moist, secluded areas. They are of ancient origin, and have not changed much during the 400 million years since the Devonian period of the Paleozoic era. They are gregarious and, because they develop from egg to adult by gradual metamorphosis, all stages and instars are found together in their preferred harborages. Cockroaches are now placed in a separate order: Dictyoptera.

These are among the most important of the household pests. In a survey of 8 large housing projects in North Carolina, 2 in each of 4 cities, cockroaches were found in 76.7 per cent of the apartments in which control work was done by project maintenance men and in 12.7 per cent of those that were regularly treated by licensed pest control operators. Cockroaches are also known to cause severe injury to plants and their fruits, particularly in greenhouses.

This group of insects is well known to biology students because, like grasshoppers, they are commonly chosen for laboratory demonstrations of a somewhat generalized type of external and internal insect morphology. They are also commonly reared for research purposes, e.g., as test insects in the investigation of new insecticides. Nearly every entomological research group continuously maintains a culture of 1 or more species of cockroaches.

Cockroaches as the Ancestors of Termites

Certain morphological, biological, and behavioural characteristics of cockroaches indicate that they were the ancestors of termites, which appeared much later in the scale of geological time. These characteristics include the possession of intestinal protozoa (flagellates) by the primitive wood-eating cockroaches, *Cryptocercus punctulatus* and *Panesthia javanica,* the former being indigenous. These protozoa convert indigestible cellulose into soluble sugar, as they do in the more primitive families of termites. Of interest in this connection is the observation that *C. punctulatus* and the termite *Zootermopsis* were found in California inhabiting the same log. Cockroaches and primitive termites also have intracellular symbionts.

This has led to the idea that the common ancestors of cockroaches and termites had both intracellular microorganisms and xylophagous flagellates. Cockroaches, except for the genus *Cryptocercus,* have retained the intracellular symbionts, and *Cryptocercus* has retained the intestinal flagellates. The primitive living termites (all the families except Termitidae), except for *Mastotermes darwiniensis* Froggatt, the most primitive of them all, have kept only the intestinal flagellates. The Termitidae have lost the xylophagous intestinal flagellates, but have acquired a rich and complex bacterial fauna.

Mastotermes darwiniensis has an egg mass similar in appearance to the egg capsule of cockroaches. Its eggs are firmly cemented together by a light-brown, gelatinous secretion that fills the interstices.

The ability many cockroaches have to mix bits of debris with mouth secretions in order to hide or disguise their oyythecae reminds one of a similar process used by termites to build nests. There are other interesting similarities.

Cryptocercus punctulatus leads a subsocial life in decayed logs, which serve as both food and shelter. An Australian cockroach, *Panesthia,* lives in distinct family groups in burrows in the soil. Each group consists of an adult male and a viviparous female, and from 10 to 20 nymphs in various instars. Soon after reaching maturity, the adults bite off their own tegmina and hindwings; wings are inconvenient for insects that inhabit burrows. This indicates the probable origin of termites' social life in subterranean galleries and their discarding of wings. The subsequent evolution of termite social organization to such an astounding degree of complexity, superficially so strikingly similar to that of the social Hymenoptera, depended on a concomitant chemical evolution of pheromones and a highly specialized chemical communication among these insects. This is a good example of the important role of pheromones in insect evolution.

The ancient cockroaches folded only a small area (anal lobe) of their hindwings, as can be seen in the fossil genus *Pycnoblattina*. The living *Mastotermes darwiniensis* also folds the anal lobe of its hindwings. The cockroaches gradually evolved more complex folding of the hindwings, while the termites, except for *M. darwiniensis*, completely lost this feature of their wing structure believed that cockroaches, walkingsticks, and termites belonged in the same insect order. She pointed out that *Cryptocercus* had xylophagous flagellates in common with primitive termites but not with other cockroaches. On the other hand, the primitive *Mastotermes darwiniensis* has intracellular bacteriocytes in common with cockroaches, but not with other termites. Also, in certain morphological features, *Cryptocercus* differs far less from *Mastotermes* than it does from many of the cockroaches.

The Place of Cockroaches in the Class Hexapoda

The place of cockroaches in the class Hexapoda has been the subject of considerable disagreement among experts during the last half-century. All the common cockroaches were formerly considered to be members of the family Blattidae of the order Orthoptera. This order included, besides the cockroaches, the grasshoppers, crickets, katydids, mantids, walkingsticks, and several other related families. In recent years, several new classifications have been proposed.

According to this classification, the cockroaches should be placed in the order Dictyoptera, which is divided into the suborders Blattaria (cockroaches) and Mantodea (mantids). Blattaria in turn is divided into the superfamilies Blattoidea and Blaberoidea. Blattoidea contains 2 families, one of which (Blattidae) includes the well-known genera *Blatta* and *Periplaneta*. Blaberoidea contains 3 families, one of which (Blattellidae) contains the genera *Blattella* and *Supella*.

One way in which cockroaches differ from the closely related grasshoppers, locusts, and crickets is that they do not have a visible ovipositor. Also, they do not lay their eggs singly or in pods, but in a hardened, purseshaped egg case called the otheca (plural, othecae). Most domestic cockroaches drop their otheca as soon as it is formed, but the German cockroach (*Blattella germanica*), generally the species most frequently found in the household, carries hers until a day or two before the eggs are ready to hatch. She is often seen with an attached otheca.

Among domiciliary cockroaches, *B. germanica* is the only species with an otheca subject to desiccation. The eggs are dependent on moisture supplied by the mother. The othecae of other species are remarkably resistant to desiccation. The otheca of *Blatta orientalis* to be covered with an oily secretion, probably derived from the cuticle of the mother. It was so effective in retarding water loss that eggs hatched in about 46 days from othecae that had been kept in an atmosphere of 0 per cent relative humidity from the age of less than 1 day.

Othecae also effectively protect the enclosed eggs from insecticides. Not until the eggs are hatched and the nymphs leave the protection of the otheca and crawl over the insecticide residue are they subjected to the effects of an insecticide treatment, provided the residue retains its efficacy that long.

World-Wide Distribution of Cockroaches

Cockroaches are tropical or subtropical in origin, and can maintain themselves outdoors in such areas. Species that became adapted to living in human habitations - "domestic" or "domiciliary" species - all originated in north or tropical Africa. Species names such as *americana, australasiae, germanica,* or *orientalis,* and the corresponding common names, are therefore misleading.

The distribution of domiciliary cockroaches throughout the world was made possible by their ability to infest the various means of intra- and intercontinental transport, particularly ships. Originally, and probably to this day, their principal means of transportation between continents has been by ships. However, transport of cockroaches in the baggage compartments and kitchens of aircraft has been abundantly documented.

For many years, the principal species infesting ships was the American cockroach, but in more recent times the German cockroach has become the dominant ship-infesting species.

To judge by the enormous numbers of cockroaches infesting ships before the advent of effective insecticides, these insects must have found abundant food and water. Even without food and water, however, cockroaches can live for considerable periods. Newly emerged adult females of the German cockroach, *Blattella germanica,* have been found to live at 27°C and 36 to 40 per cent relative humidity for an average of 20.1 days without food or water, 14.7 days with dry food (dog biscuits) and no water, 35.1 days with water but no food, and 82 days with both food and water. The corresponding figures for the American cockroach, *Periplaneta americana,* were 41.7, 40.1, 89.6, and 190. The apparent ability of the 2 species to survive slightly longer without food and water than with dry food alone is interesting, but in any case, cockroaches are well adapted to dispersal over great distances by man, even when food and water are lacking.

Domestic species of cockroaches never or rarely fly, but are readily carried about in and on such items as sacks, cartons, or packages (particularly corrugated cartons) of food, in laundry, or in kitchen appliances and furniture. Beverage cartons are particularly important means of distributing cockroaches. The cartons are often contaminated with spilled syrups or malts, which attract cockroaches. Empty and unrinsed softdrink or beer bottles in night clubs, restaurants, markets, and homes form a part of the infestation chain. More than 200 cockroach nymphs have been found in a single soft-drink bottle brought to a market for exchange.

Cockroaches may be found by the thousands in insulation in the walls of refrigerators and gas or electric ranges, surviving for months if necessary without access to their usual foods and feeding only on cast skins and dead insects. Cockroaches may become established in basements or crawl spaces, particularly, if these places are dark and damp, and they may then enter the building around utility pipes, air ducts or ventilators, or under doors.

Economic Importance of Cockroaches

Cockroaches rank with termites as the most important insects that must be dealt with by the owners or occupants of homes or other buildings or by pest control operators hired to combat insects. Flies and mosquitoes may be more important as household pests in some areas, but they breed elsewhere and then invade dwellings. Therefore, in the United States and many other countries, fly and mosquito control is generally considered to be primarily a community or regional responsibility, and the homeowner is generally left with the job of controlling only minor or seasonal infestation.

The Cockroach as a Domestic Nuisance

Cockroaches breed within the home, sharing with man his food, water, shelter, and warmth, and are thus active throughout the year. They consume any human or animal food or beverage, as well as dead animals and plant materials, leather, glue, hair, wallpaper, fabrics, and the starch in bookbindings. They contaminate food, imparting an unpleasant odor and taste, and may transmit disease organisms.

Although cockroaches are nocturnal, if they are abundant a few of them may be seen during daylight hours, particularly if articles in pantries, cabinets, and closets are moved about to disturb their hiding places. Other evidences of a cockroach infestation may be dead cockroaches, cast skins of the various nymphal instars, empty egg cases, and fecal droppings. The droppings are variable in size, and range from as small as "flyspecks" to as large as mouse droppings. Intersections and corners of shelves and the hinges of cabinet, pantry, and even closet doors show fecal stains when there are heavy infestations. The musty odor left on objects cockroaches contact may also be evident in heavy infestations.

Cockroaches as Disease Carriers

Many of the organisms causing diseases of man have been found on the tarsi and other body parts, or in the fecal pellets, of cockroaches, but these insects are most commonly implicated in the transmission of *Salmonella,* the causal agent of food poisoning. When the feces of American cockroaches infected with *Salmonella oranienburg* were spread on human foods and on glass, the bacteria survived for the following periods: on corn flakes, 3.25 years; on crackers, over 4.25 years; and on glass slides, over 3.67 years. Mice placed in jars containing minute quantities of infected cockroach feces became themselves infected in as brief a period as 1 day.

Parasitic toxoplasmosis is a disease that is believed to infect more than a third of all adult Americans at some time during their lives. It is usually a mild disease, but can be extremely serious in pregnant women, causing congenital defects in an unborn child while remaining mild or symptomless in the mother. The protozoan parasite, *Toxoplasma,* can be acquired by eating infected raw or undercooked meat. If the blood serum of an infected person is added to cultures of the parasites, the differing colours of fluorescence of the latter will indicate either a positive or negative reaction and thereby serve as a test for the disease.

Cats that consume birds or rodents carrying the parasite can become carriers. Filth flies have been found capable of transmitting infectious *Toxoplasma* cysts to human food 1 or 2 days after consuming infected cat feces. In laboratory tests, *Toxoplasma* parasites were isolated from the digestive tracts and feces of cockroaches as long as 7 and 10 days, respectively, after the cockroaches had last been in contact with infected cat feces. Filth flies and cockroaches might consume infectious cat feces and contaminate human food. They might also serve as food for birds and small rodents that could in turn be eaten by cats.

Goodwin recorded the successful passage of hepatitis B antigen (AB-Ag) through the alimentary tracts of American cockroaches, and identified the antigen in the feces of the insects. The hepatitis antigen was found in the feces of the cockroaches for up to 9 days after their initial exposure to the antigen-positive test meal.

COCKROACH BITES

Cockroaches biting man, such instances are rare and not serious; they are discussed here only because questions are often asked concerning the biting ability of cockroaches. There have been a number of reports in the literature of cockroaches being so numerous in ships that they gnawed the skin and nails of the men on board; in one case, the sailors wore gloves while sleeping to prevent hordes of the insects from gnawing off their fingernails. Cockroaches are most likely to gnaw fingernails, eyelashes, and the callused portions of hands and feet, but can also cause small wounds on softer skin. *Periplaneta americana* and *P. australasiae* are the species most often implicated.

The Cockroach Allergen

Some people are allergic to cockroaches. In one investigation among 253 normal persons, 7.5 per cent showed positive skin tests with extracts of cockroaches compared with 28 per cent of an unselected group of 114 allergic patients. Skin-sensitizing antibodies were present in the blood sera of positive reactors. The allergen is thermostabile. There is evidence that in some instances symptoms attributable to food allergy may be caused by cockroach allergen with which the food is contaminated.

In an investigation to test the age of onset of skin reactivity, 38 out of 102 allergic children, ranging from infants to 12 years old, gave positive cutaneous reactions to body extracts of the German cockroach, *Blattella germanica*, compared with only 5 out of 100 nonallergic children. A 4-year-old asthmatic child was the youngest to give a positive reaction. In a later investigation, it was found that an extract of *B. germanica* caused an attack of asthma in 10 asthmatic persons with skin hypersensitivity to the extract and other allergens, but not in asthmatics without such skin hypersensitivity.

An allergen in the feces of *B. germanica* acts as an ingestant when it contaminates food and as an inhalant when dried fecal particles become incorporated with house dust.

Natural Enemies of Cockroaches

Species from at least 6 families of Hymenoptera had been recorded as developing on cockroach eggs. They quoted Edmonds as stating that evaniid egg parasites were so abundant in a home that the occupants considered them a nuisance, although the oriental cockroaches in the basement did not annoy them. Roth and Willis also listed many predators, including scorpions, spiders, dragonflies, mantids, bugs, beetles, wasps, ants, toads, frogs, lizards, birds, poultry, and various mammals. Cannibalism among cockroaches, including the devouring of othecae, has been noted by many investigators even when food was adequate.

When humidity is too high in cockroach culture colonies, mites may become numerous, and have been known to cause German cockroaches to drop their othecae prematurely, resulting in a low percentage of eggs hatching. Control of mites with acaricides has been known to increase the vigour of cockroach colonie. A pterygosomid mite *Pimeliaphilus podapolipophagus* Trgrdh, must actually feed on live cockroaches to survive; it cannot subsist on feces, cast skins, or dead insects. It has been accused of biting people, and its presence in homes is an indication of cockroach infestation.

Rearing Cockroaches Experimentally

A continuous supply of several species of cockroaches has been reared for experimental purposes in our laboratory at UCLA. These species are, in the order of what we consider to be their importance in California, the German, oriental, brownbanded, and American cockroach. Later the Australian and the brown cockroach were added.

The insects were originally reared in 15 gal (57-L) garbage cans with an electric barrier, later replaced by 20-gal (76-L) plastic trash cans. A double layer of strips of aluminum foil with a strip of insulation tape between them, and with the terminals from a 22.5-volt battery connected to the aluminum strips, serves as an electrified barrier to the escape of the cockroaches. In each trash can there is a poultry-watering device, Purrina Dog Chow, and 3 mazes of concentrically rolled and closely approximate layers of corrugated paper to serve as the harborage areas that cockroaches require for optimum development. The trash cans are covered with their usual plastic lids, provided with screens for ventilation.

It is important to prevent contamination of a colony by stray individuals of other species. It is especially important to keep German cockroaches out of colonies of other species because of the ability of this insect to thrive and rapidly increase its numbers at the expense of the others.

In the cockroach rearing room, boric acid powder applied under benches, radiators, appliances, cabinets, etc., will serve to permanently prevent a buildup of the cockroach population, in the rearing room and adjacent rooms, from those insects which escape when they are being removed for experimental work.

Twenty cans of German cockroaches and a can or more of each of the other 5 species are continuously maintained in our laboratory. The temperature is kept near 80 F in the rearing room. Provided that a water supply is constantly maintained, the author and his colleagues have had no indication that the maintenance of relative humidity above ambient is of any advantage in the rearing of cockroaches, at least in the rearing cans just described. With ambient humidity and monthly removal of the debris at the bottoms of the cans, infestation of German cockroaches by parasitic mites has been avoided. American cockroaches drop their egg cases to the bottoms of their rearing cans; therefore, the cans are vacuumed only about twice a year, but their colonies do not produce so much debris and do not seem to be so susceptible to mite infestation as those of the German cockroaches.

German Cockroach

The German cockroach is generally the most important cockroach species in the United States and many other areas of the world in homes, apartment buildings, restaurants, and in the food preparation and food-storage rooms of hotels, hospitals, and other private and public buildings in which food is prepared, stored, or served.

Description. Adult German cockroaches are 1.3 to 1.6 cm long, pale brown or tan, and have 2 parallel dark streaks on the pronotum. They have chewing mouthparts. Their movements are very rapid when they are disturbed. They are nocturnal. If a few are seen crawling about in open spaces during daylight hours, this indicates that the infestation is already severe. The female is darker and has a broader abdomen than the male, and is rounded posteriorly. If German cockroaches are placed on their backs, the males are easily distinguished by the yellowish, slender abdomen, tapering gradually to the posterior end. The abdominal cerci of the males have 11 segments and those of the females have 12.

In their investigation of the German cockroach, the ratio of females to males to be 1.12:1. The male of the German cockroach could not detect the female from a distance, even when in close proximity, but had to make physical contact, ordinarily via the antennae. The antennae of the female, as well as other body regions, contain a chloroform-soluble, nonvolatile substance that will stimulate the male sexually. Thus, sex discrimination by males is mainly owing to "contact chemoreception." It follows that the male German cockroach probably cannot be attracted from a distance by means of synthetic sex attractants (provided they are available) as some other insect species can be.

When male and female meet, their antennae touch and vibrate against each other. The male then turns around, raises his wings to expose the orifices of a pair of dorsal glands located on the seventh and eighth tergites, and extends his abdominal segments to expose the openings of the 2 pairs of glands. The glands are not normally visible, being covered by the wings and by the margins of the preceding abdominal sclerites. Both wing-raising and extension of the abdomen are required to uncover the glands. The female eats a secretion from these glands. After she has fed for a few seconds, the male-pushes his abdomen farther back, and connection of the genitalia is made. The male then moves out from under the female, and the pair remain attached in a linear position for an average of about 86 minutes.

These authors also pointed out that the fact that German cockroaches do not need to be attracted from great distances, as do many other insects for the meeting of sexes, is understandable in view of the habits of cockroaches in general. The chance meeting of opposite sexes is enhanced by the fact that they are negatively phototactic and positively thigmotactic (principally guided by contact), and that they are gregarious, with large numbers seeking the same environment. They seek secluded and particularly very narrow hiding places, such as cracks, crevices, and voids, especially those having optimum temperature and moisture conditions. Attraction to the odor of the species is another factor that favours aggregation. Once the sexes have been brought together by these different stimuli, more refined stimuli lead to sexual discrimination.

Life Cycle. The German cockroach lays more eggs and has more generations per year than the other common domiciliary species, and therefore tends to increase in numbers much more rapidly. There may be 3 or 4 generations per year. In one investigation at a temperature of 76 F, the developmental period varied from 54 to 215 days, with an average of 103. It has been found to be as little as 60 days for males and 65 days for females at 95 F and 90 to 95 per cent relative humidity. An individual German cockroach, however, lives an average of 200 days at room temperature and may live as long as 10 months. A comparison of the developmental periods of the 4 principal domiciliary cockroaches at 82 F and at ordinary room temperature.

The Otheca (Egg Capsule). The German cockroach female produces her first otheca 11 or 12 days after becoming an adult. Ohecae can be produced parthenogenetically, but although embryos develop, the eggs do not hatch. Unlike other domiciliary species, the German cockroach female carries her otheca for as long as a month, until a day or two before the eggs are ready to hatch, then drops it anywhere. The eggs occasionally hatch while the otheca is still attached. The female generally deposits 4 or 5 othecae, but there can be as many as 8. The females of other cockroach species drop the otheca as soon as it is fully formed. Thus, the abundance of females with their attached, brownish, purseshaped othecae is a characteristic of a German cockroach infestation.

The otheca is about 8 x 3 mm-rather large in relation to the size of the female. As with all common cockroach species, the eggs are located in 2 parallel rows. The locations of the eggs are indicated by corresponding divisions on the outer wall of the brownish otheca. When the embryos have developed to the point where they exert enough pressure, the otheca splits open at the top and the nymphs wriggle out. The female will eat the young nymphs if no water is available to her.

There are usually 30 to 40 eggs per otheca, but there can be as many as 48. In one investigation, an average of 29.9 nymphs were hatched per otheca in an average of about 28 days at ordinary room temperature, while in another investigation at 95 F, eggs hatched in 14 days. After the fourth otheca is produced, the number of eggs per otheca gradually decreases to about 75 per centof the original number in the seventh and eighth othecae.

The eggs become desiccated if the otheca is removed from the female any time before its normal time to drop, for they must obtain moisture from the body of the mother. The end of the otheca that is attached to the female is relatively soft and permeable to water, and is not so heavily sclerotized as the posterior end. German cockroach othecae placed with their anterior ends on wet filter paper gained weight, whereas those with their posterior ends on the same wet paper lost weight, even though the humidity was high in the covered petri dish in which the othecae were kept. Probably, the wall of the otheca in contact with the female's genital pouch is permeable to water.

Even before the otheca begins to emerge, the abdomen of the female is greatly distended. Once the translucent tip of the otheca becomes visible, the entire otheca will be fully developed and entirely visible by the following day, changing from white to pink within a few hours. Within a day or two, it becomes light brown and finally chestnut. It is turned with the keel to the left or right. German cockroach nymphs have 6 or 7 instars. As in the case of all insects with gradual or incomplete metamorphosis, there is no abrupt change in appearance between immature and adult forms, except that the adults have wings. Nevertheless, the layman will not necessarily associate the tiny, newly hatched German cockroaches or even the more mature nymphs with the adult insects unless he sees all instars and stages together in considerable numbers. The smaller nymphal instars are sometimes isolated from the remainder of the colony, particularly when they have gained access to a crack or crevice too narrow for the older nymphs and adults.

The first-instar nymph is only 3 mm long. The body is dark gray to almost black, except for the second and third thoracic segments, which are pale brown. The pale-brown band conspicuously characterizes the first-instar nymph. In succeeding nymphal instars, the light band becomes narrower and extends in both directions to become a median longitudinal stripe. In its anterior extension, it eventually becomes the median pale-brown stripe dividing the "two parallel dark streaks" (already mentioned) that characterize the pronotum of the adult. The remainder of this stripe is covered by the wings.

Completely white nymphs or adults may be seen in a cockroach colony. These are newly molted individuals that have not yet had time to harden their cuticles and to acquire the normal colour for the species.

A high degree of mortality occurs during the molting periods. When it was 40 per cent. He found that about half the insects died of natural causes before reaching maturity. Molted skins were quickly eaten by the nymphs that emerged from them or by other cockroaches that happened to be near-by.

Aggregation Pheromone

As already stated, because cockroaches have a gradual metamorphosis, all immature instars may be seen together with the adults. They tend to congregate in a single area, leaving it to search for food and water, then returning. An aggregation pheromone, present in the feces and on the bodies of the insects (it spreads easily on the greasy epicuticular lipid), is responsible for this tendency to aggregate. The gregarious behaviour of nymphs favours their growth and development; they do not develop so well when isolated. Isolated nymphs that could only see or smell the others, or received contact only with the washings of the body surfaces of other nymphs, or with filter paper contaminated with feces, did not grow at the group rate. This indicated that a "psychological factor" affecting the endocrine system by means of tactile stimulation, chiefly via the antennae, was the principal factor involved in growth acceleration.

Effect of Excessive Crowding. The author investigated the effect of crowding in a "choice box," 30 cm square and 10 cm high, with a 12-mm hole at the top of the partition wall between the 2 halves. When there are only 20 to 40 cockroaches in the choice box, most of them will spend nearly all their time in the dark half, for they are negatively phototropic. The insects tend to congregate at the intersection of 2 plane surfaces or in corners. When such areas are filled, the insects will gather in small, scattered groups on plane surfaces. However, despite the aggregation pheromone, German cockroaches have some aversion to being crowded too closely together. They prefer to leave some space between themselves and their neighbours. They sometimes initially display aggressive behaviour toward their neighbours when actual physical contact is made. However, if large numbers are forced to occupy limited space, they soon appear to get used to this situation.

When 200 adult male German cockroaches were placed in a choice box, only 11 per cent were in the light half 3 days later, compared with an average of 16.2 per cent in the light halves of 5 choice boxes, each containing 20 adult male cockroaches. On the other hand,, when 959 adult German cockroaches of mixed sexes and 530 nymphs were placed in the dark half of a choice box, they were so crowded that they made physical contact with one another, and in some cases were one on top of the other. They were left that way for 2 hours, and then a cork occluding the 12-mm hole in the partition wall was removed. Large numbers of cockroaches left the dark compartment. On the third day of the experiment, the insects were anesthetized and,counted. Of the live insects, 36.1 per cent of the adults and 41.7 per cent of the nymphs were in the light compartment of the choice box, preferring the light to an excessively crowded condition in the dark. If German cockroaches are found in parts of an untreated building far removed from food and water or from preferred habitats, it is an indication of excessive crowding in their customary harborages.

Crowding in Narrow Spaces. Adult German cockroaches can move about in space only 1.6 mm in width or depth. When given a choice among 8 spaces between masonite plaques placed one above the other and ranging in distances from 1.6 to 12.7 mm apart, in 1.6-mm increments, 67 per cent of adult German cockroaches gathered in the 4.8-mm space. When the plaques were vertically arranged, this percentage increased to 85. Only a few adults gathered in the 1.6-mm space. In our laboratory we found that only the first 3 instars of a population of all ages of German cockroaches crawled from a light to a dark compartment of a 5-cm-wide wooden box for a distance of 2 cm under a strip of masonite that was only 1 mm above the floor of the box. Only 80 per cent of the adult population (6 males and 2 females without othecae) crawled through a 2-mm aperture in a 24-hour period, even though the insects were driven by hunger and negative phototropism to seek food and darkness. No adults were able to crawl from the light to the dark compartment when the 2-mm aperture was between the wooden partition wall (1 cm thick) and a masonite ceiling.

The author once observed a severe infestation of German cockroaches in the food-storage room of a large "rest home." Leaking or spilled packages of flour, cereal, or similar foods on the shelves provided plentiful food for the insects. Cockroaches were crowded into a space of 2 or 3 mm between a shelf and the wall. They were completely hidden from view, except for a continuous band of waving antennae that extended from the aperture, and could be seen by looking beneath the shelf.

In the living space of a house, German cockroaches are most abundant in the kitchen, and this is also true of other domiciliary species. During the day they may be found under stoves or ranges and refrigerators, and in the insulation in the walls of appliances; under the sink and especially in the "dead space" between the sink and the wall; under and behind the water heater, particularly if it is enclosed in a cabinet; in, under, or behind cabinets, pantries, and closets; behind baseboards and moldings; and in other cracks, crevices, and dark, protected areas. They venture forth at night in search of food and water.

German cockroaches are also likely to be found in the bathroom, particularly if the kitchen and bathroom share a common wall, for then they can infest a common wall void and pass from one room to the other via areas surrounding utility pipes or the louvers and razor-blade receptacle of the built-in medicine cabinet. In moderate to severe infestations, they can be found throughout the house in protected areas, such as closets, dressers, sofas, electrical appliances such as radios and television sets, under stairwells, behind moldings, picture frames, etc. All such places must be thoroughly treated with insecticide for effective control.

Television sets in particular are becoming increasingly important harborages for both German and brownbanded cockroaches. Not only the warmth, darkness, and the edible glue used in construction, but often the bits of human food commonly found in the vicinity of television sets are attractive to cockroaches and combine to provide an ideal harborage.

In severe infestations, enormous numbers of German cockroaches may sometimes be found in the attic and in wall voids. They use the wall voids to move either laterally or vertically from one room to another or from one apartment to another. They may also be found in the crawl space under the house, at the bases of foundation walls, in cracks in the sidewalk, in the lawn and under shrubbery, and in outdoor incinerators or garbage-disposal bins.

German Cockroaches Finding New Habitats

Before German cockroaches became almost universally resistant to chlordane in the 1950's, they were known to inhabit kitchen areas almost exclusively. Since that time, they have extended their habitats until they are now often distributed widely throughout the house, at least in the cases of more severe infestations. This is a characteristic formerly attributed only to the brownbanded cockroach, *Supella longipalpa* (F.). It may be that because of the switch from chlordane to organophosphorus and carbamate insecticides, German cockroaches were not only more strongly repelled by the latter insecticides, which was easily demonstrated, but that they were more successful in surviving initial contacts with the toxicants, and found new harborages free of insecticides in a wide variety of locations throughout the infested premises.

This new behaviour might be inherited because of its survival value.

There are other reasons for the increasing dispersal of German cockroaches throughout a house that arise from changes in furnishings and in man's living habits. Central heating has become common, resulting in temperatures in all areas that are conducive to cockroach breeding, even in winter. Even if some rooms are cooler at night, the insects can find warmth in such appliances as electric clocks, radios, or television sets.

Some people eat their meals, or at least light snacks, near the television, which could be in the livingroom, bedroom, or den. A few crumbs, pieces of cookies, etc., can sustain many cockroaches. Many modern homes also provide more sources of water than were formerly found in homes. There may be 2 or 3 bathrooms, as well as air-conditioners with evaporating pans. Beer cans and soft-drink bottles not completely emptied may be left in more locations, now that homes are more extensively utilized than they formerly were, and these also provide sources of both food and water.

That Influence Aggregation

The earliest insects had sheltered and restricted habitats that were at least reasonably damp, such as beneath stones and litter or under loose bark, as one might expect on the basis of the relatively poor waterproofing mechanism of their cuticles. Ametabolous or paurometabolous development made a continuously cryptobiotic life style possible. Food and mates were found by chance encounters. It was improved cuticular waterproofing and the development of wings that freed insects from a restricted environment. An improved olfactory sense became advantageous. Eventually, the pheromones and sophisticated chemical communication systems of some of the modern insects were evolved.

Among household pests, the silverfish and cockroaches are among the cryptobiotic insects. Although most cockroaches have wings, they use them sparingly or not at all. Accordingly, one might expect that neither silverfish nor cockroaches would be effectively attracted to foods or chemical lures, at least not over distances that might lead to effective chemical trap devices, and this has indeed been found to be true for both these relatively primitive insect orders.

In the case of *Blattella germanica* and *Blatta orientalis*, the species aggregation odor was shown to be largely responsible for their gregarious behaviour, but the stimulus was effective for only a short distance. As stated earlier, sex stimulus, although of chemical origin, is limited in cockroaches and requires actual physical contact. Even food odors are only feebly attractive, and will attract German cockroaches only if they occur in areas normally frequented by the insects or in the routes in which they would normally travel. For example, in a room, cockroaches normally travel mainly along wall intersections, the intersections of the floor and the walls, the intersections of the walls and the ceiling, the intersections of shelves and pantry or cabinet walls, and infrequently in other areas.

When experimenting with German cockroaches, that in a kitchen area in which food was available to the insects, an average of 2.45 per baited trap-jar was trapped in jars placed along the baseboards, but none in jars placed in the centre df the kitchen floor. When all other food was removed from the kitchen, an average of 13.5 insects was trapped in the baited jars placed along the baseboard and 1.15 in the centre of the room.

Hunger is one of the factors influencing exploratory activity of cockroaches, but the physical features of a habitat are the principal factors influencing the areas in which the insects will travel.

Factors such as pheromones and food odors can become operative only after these physical features have already attracted the insects. For insecticides or even poison baits to be effective, they must be applied in areas in which cockroaches would normally travel or congregate.

Mass Migratioms of German Cockroaches

These mass migrations have been known to occur without the stimulus of insecticide treatment. For example, a "vast army" of German cockroaches leaving a restaurant on a "dark, drizzly day" and crossing a muddy street to other buildings in a migration that lasted for 3 hours. The migrating insects were nearly all females with othecae. No unusual activity, such as cleaning or insecticide treatment, had taken place in the restaurant. Dr. Howard concluded that this mass movement of cockroaches was a true migration occasioned by overcrowding in the restaurant. Another instance of active migration of German cockroaches from a restaurant to another building was reported by Mielke.

Brownbanded Cockroach

Description. The adult brownbanded cockroach is approximately the same size. as the German cockroach. The male is 13 to 14.5 mm long, and the female slightly shorter. The abdomen of the female is much broader and more rounded posteriorly than that of the male. The wings of the male cover the abdomen completely, while the wings of the female are short, and never cover the entire abdomen. Males may often be seen flying when a colony is disturbed, but the females cannot fly.

From the dorsal aspect, the adults are dark brown, particularly the head and thorax and the exposed segments of the female, which may in fact range to black. The wings of the female are reddish brown to very dark brown throughout, while those of the male are dark brown at the base, becoming increasingly lighter posteriorly. Both sexes have a band of pale brown at the base of the wings and another band of pale brown a third of the distance from the base, which gives the adults a "banded" appearance. The wings and body of the female are much darker than those of the male. The ventral body surface and legs of the male and the legs of the female are light strawcoloured. The lateral margins of the pronotum of both sexes are translucent, and this is also true of a narrower lateral margin of the anterior portion of the wing. The ratio of females to males was found in one investigation to be 1.32:1.

Life Cycle. The developmental period of the brownbanded cockroach is considerably longer than that of the German cockroach. The temperature for optimum development is above 27° C, and homes are seldom kept at such high temperatures. Therefore, the biotic potential for the brownbanded cockroach is considerably less than for the German cockroach. In California, it is found in only about 51 per cent as many homes as the Geiman cockroach, but where food is abundant and temperatures are high, the species can become extremely plentiful.

The Otheca: The females carry their capsules for 24 to 36 hours before attaching them to some object, possibly the kitchen sink, furniture, walls, shelves, bedding, draperies, behind pictures on walls, or other handy places, usually in clusters corners under chairs and tables are among the favoured places. Individual clusters on the walls of rearing cans in the author's laboratory have at times contained 50 or more egg capsules. One housewife, in attempting to describe them, said they appeared to be "clusters of seeds". This habit of attaching the othecae to furniture and other household objects results in the species being readily disseminated over great distances. The otheca of the brownbanded cockroach measures about 4 x 2.5 mm-the smallest of the common domiciliary cockroach othecae. It varies in colour from yellowish to reddish brown. A few days after it is formed, the fertile eggs show greenish through the walls of the otheca, and shortly before hatching, the eyes of the young are visible.

The incubation period for the eggs of the brownbanded cockroach was found to be the longest among 6 cockroach species investigated. It averaged 49 days at 28°C and 95.6 days at 22.5° C. About 14 fertile othecae were formed per female.

The maximum number of eggs normally produced per capsule was 18, but the average number hatching was 13.2.

Nymphal Instars. From the dorsal aspect, the nymphs are also basically dark brown in colour, but with a narrow, pale margin on the thorax. The meso- and metathorax are pale brown, and the abdomen has a broad, pale area. The dark and pale areas of the nymphs are in striking contrast, and easily identify this species.

There are 6 to 8 nymphal instars. At room temperature, the average period for nymphal development to be 161 days for males and 162 days for females, but at about 29°C, the average periods were 90 and 95 days, respectively. At room temperature, extremes of developmental periods were 95 and 276 days. The length of adult life varied from 131 to 315 days, with an average of 206 days.

Habits. The brownbanded cockroach was once believed to be unique among domiciliary cockroaches in its tendency to distribute itself throughout a house, even in bedrooms, furniture, and closets, particularly on high shelves, behind pictures on walls, and behind moldings. "Except when in search of food, it seldom visits the kitchen, and confines its activities to other parts of the house." About the German cockroach, they stated: "In an infested home, it is confined to the kitchen and lavatories, where it hides behind baseboards, in cupboards, iceboxes, and dark corners, and around water pipes." Apparently, the habits of the 2 species have changed, for now there appears to be no great difference in the way they are distributed in a house or apartment.

The change in the habits of the German cockroach coincided with widespread resistance to chlordane, which in the years following World War II was almost universally used for control of this species. Both species are now most abundant in the kitchen areas, and either can be widely distributed throughout a house or apartment. Sometimes, mixed populations of the 2 species are seen. However, as stated earlier, infestations of brownbanded cockroaches generally occur only in houses or apartments in which temperatures are higher than most people would consider to be comfortable.

Oriental Cockroach

In common with the other important domiciliary cockroaches, the oriental cockroach, despite its name, is believed to have originated in north Africa in areas which combine summer heat and moderate winters. Although in some parts of the world, as in Great Britain, the oriental cockroach is by far the most important cockroach pest, the German cockroach is much more important in the United States.

In California, the oriental cockroach is found in homes more frequently than the brownbanded cockroach, but does not reach such large numbers or become so widely distributed in the house. Nevertheless, pest control operators report more calls for control of the oriental than of the brownbanded cockroach. Its large size and black colour cause it to be regarded as a particularly repulsive pest. The housewife prefers to believe that the insect is a "water bug" or "black beetle" rather than a cockroach, but usually loses no time in calling for the exterminator.

Description. The adult females of the oriental cockroach are about 3 cm long. The male is about 2.5 cm long, and is much narrower than the female. Both sexes vary from dark reddish brown to black, usually the latter. The female has small, rudimentary, functionless wings, and the male, although it has well-developed wings covering about 75 per cent of its abdomen, does not fly. The tarsi of both nymphs and adults lack an arolium, and there they cannot climb smooth, vertical surfaces. This species is relatively sluggish in its movements. The ratio of females to males has been found to be 1: 1.15.

Life Cycle. The developmental period of the oriental cockroach is more than a year, even under indoor conditions, and for some individuals it may be prolonged to about 2 years.

The insect may also live for long periods after reaching maturity. At room temperature, the length of adult lives of 5 females varied from 34 to. 181 days. Among 4 males, the length of adult lives varied from 112 to 160 days. Whereas it may require a long time for an appreciable infestation of oriental cockroaches to develop, enormous populations are eventually possible. Ordinarily, the oriental cockroach does not become so abundant as the German cockroach in buildings, but can become very numerous in secluded places, such as basements or crawl spaces, if they are damp. The species thrives better outdoors than other domiciliary cockroaches in the United States, even under quite severe climatic conditions.

The Otheca. The female carries her otheca for about 30 hours after it is formed and filled with eggs. It is dropped, or attached to debris or food material, in some protected place. It is large (10 x 5 mm), and when first deposited it is reddish brown and soft, but becomes black, hard, and brittle. Although the number of othecae.formed varies greatly, in an observation of 8 females, the average number was 8. A perfect capsule contained 16 eggs. Anothcr investigator observed that females produced from to 4 othecae, with the usual number being 2. The incubation period was noted to vary from 42 days at 29.5 °Cto 81 days at 21 °C (85 and 70 °F), and was approximately 2 months at the usual room temperatures.

Brown Cockroach

This species closely resembles the American cockroach in habits and appearance, but is more confined to the tropics in its geographic distribution. It is common in the southern and southeastern United States from the Carolinas to Florida and westward to eastern Texas. It has been collected indoors at points farther north. The apartment was occupied by tenants who had arrived 4 months earlier from San Antonio, Texas, where the predominant species of *Periplaneta* is *P. brunnea.* It is also known that the brown cockroach is more common than the American cockroach in certain other areas of the South. Another accidental introduction of brown cockroaches from Texas was noted.

Despite the close resemblance of *P. brunnea* and P. americana, hybridization does not occur. A mixed colony of these 2 species plus *P. fuliginosa* was maintained for more than 3 years without the production of any recognizable hybrids.

Description of Adults. The adults are dark reddish brown, resembling the American cockroach, but are somewhat darker in colour. The 2 species also resemble each other in the yellowish margin of the pronotum. When measuring 25 individuals of each sex among those reared in our laboratory, the lengths and widths of *Periplaneta americana, P. brunnea,* and *P. australasiae* were recorded, and are shown in table 4.

The data show that the American and brown cockroach are similar in length, but that the latter is the wider of the 2 species. The wings of the male brown cockroach do not extend beyond the tip of the abdomen as far as those of the American cockroach. Both the brown and the American cockroaches can be distinguished from the Australian cockroach by their slightly larger size and the absence of the yellowish streak on the outer edge of the base of each forewing that so distinctly characterizes *P. australasiae.*

The cercus of the American cockroach is stout basally, and tapers markedly toward the tip. The sides of the last segment are more or less parallel, and 2 or more times as long as the basal width. The cercus of the brown cockroach is also stout, but the appendage is more evenly spindleshaped, with the last segment somewhat triangular and less than twice as long as its basal width. *Periplaneta brunnea* can fly, but its flight is usually of the gliding type.

Life Cycle. Judging only from the enormous production of *P. brunnea* in our rearing room at normal room temperature when compared with other species of *Periplaneta,* it appears to have an alarming reproductive potential. Possibly. there are ecological factors under actual field conditions that have thus far resulted in this species being generally less abundant than the American cockroach, even in the South.

Copulation occurred within a few hours after a female completed development, and egg deposition began 16 to 20 days later. It continued throughout her life. The periods between oýÿtheca depositions varied greatly, but could be as little as 5 to 6 days.

Other Cockroach Species

Many people familiar with the common domiciliary species of the United States are not aware of the large number of other species, indigenous or accidentally introduced, that occur in this country. Some are at times abundant in certain regions.

Pennsylvania Wood Cockroach

This cockroach is indigenous to the United States, ranging through the midwestern, southern, and eastern states and north into Canada. It as "a trim, pretty creature, with its chestnut-brown colour, and thorax and wing pads edged in white," but the variability of the species in different areas, and did not consider a general description of colour to be of any value. The males are fully winged, and can occasionally be seen flying about outdoors. The wings of the females are short and nonfunctional. The males are considerably larger than the females, being 2.2 to 3 cm long compared with 1.3 to 2 cm for the females. Houses in rural districts and near wooded areas are commonly invaded, particularly by the males, which fly and are attracted to lights. This species is also frequently seen by motorists when driving through the woods at dusk.

Parcoblatta pennsylvanica happens to have the widest distribution, but there are at least a dozen other species of the genus that occasionally find their way into houses in the United States, particularly *P. lata* (Brunner) and *P. virginica* (Brunner). Species of *Parcoblatta* are found throughout the country. They are generally seen in wooded areas, but one, *P. desertae,* is found only in desert and semidesert mountains of the Southwest. *Parcoblatta americana* (Scudder) is the only species of the genus found in California. It is small, the males being 12 to 14 mm long and the females 9 to 13 mm. The females are wingless. The colour is generally shining buff or brown, ranging to black in some females in the region of Mount Shasta, California. The smaller and paler specimens are found in the more arid and barren regions.

Smoky Brown Cockroach

This species is uniformly shining brownish black or mahogany in colour. Both sexes have wings that extend beyond the tip of the abdomen. It is common in the southern states from central Texas eastward, and has been found in heated buildings in isolated locations as far north as Iowa, Indiana, and Illinois. This is the only uniformly dark-coloured species of *Periplaneta* in North America, and is also the only species not known to occur in other parts of the world. It flies to lights at night, and is sometimes attracted to lighted porches. It has been recorded as second only to the German cockroach as a pest in houses in southwest Georgia, but is more commonly found in garages, outbuildings, and woodpiles. It is sometimes a pest in greenhouses.

Control of Cockroaches

Good Housekeeping. Cockroaches cannot live in human habitations without the food and water they provide. Obviously, good housekeeping is an important factor in control. Food particles should not be allowed to remain on shelves or floors. Dishes should not be left unwashed after a meal, particularly overnight. Areas under refrigerators, stoves, sinks, and furniture should be checked to see whether food has accidentally fallen or has been swept into them or dropped there by children or pets. If pets are fed indoors, food particles should not be left in their dishes after feeding, particularly not overnight. Keep cereals, crackers, meats, vegetables, flour, pet foods, and all other foods covered or in a refrigerator at all times between uses. Keep garbage and waste containers covered, or empty them into a covered container every night.

Water is often available as condensation or spillage under the refrigerator. Patches of water often remain for hours in certain locations after a floor mopping or following the use of a washing machine. Other common sources of water for cockroaches are the traps of sinks, washbasins, tubs' and toilet bowls; flush tanks; condensation on cold pipes and windows; leaking pipes and faucets; miscellaneous water-filled containers such as pets' drinking dishes, aquaria, vases, empty beverage bottles; and the juices from soft fruits and vegetables.

Cockroach infestation is favoured by a cluttered household. Stored magazines and newspapers, corrugated-paper boxes and piles of paper bags should be eliminated. They not only provide harborage and breeding areas for cockroaches, but also provide areas into which the cockroaches can escape from insecticide residues when control is attempted. Insecticides are repellent to cockroaches in varying degrees.

Application of Insecticides. If liquid sprays are to be applied, and ideally for dusts also, all shelves and drawers should be emptied and cleaned. After treatment, shelf paper may be placed on sprayed areas and on dust deposits. If a relatively safe dust such as boric acid is to be used, it is recommended that the operator reach carefully behind shelf contents and apply the dust with a bellows powder blower ("Getz Gun"), bulb duster, or plastic squeeze bottle in a narrow band along intersections and into corners and crevices. Shelf paper lifted while the powder is applied may then be allowed to fall back into place.

When applying insecticides for cockroach control, pest control operators ordinarily ask the occupants of the treated premises to remove cooking and eating utensils and foodstuffs when there is danger of their becoming contaminated with insecticide. For example, they may be placed on a table in the centre of a room and covered with a sheet. Occupants are asked to leave the building and remove their pets during treatment.

Treatment for cockroaches, like so much of household pest control, is "spot treatment." The insecticide is applied to relatively limited areas where cockroaches are likely to hide during the day or where they are likely to travel at night. Most cockroaches congregate in dark, warm. damp places, such as under and behind the refrigerator and stove and under the sink, particularly the "dead space" between the sink and the wall. The most thorough treatment can be made under the refrigerator and stove if the kick panels are first removed, allowing the person applying the spray or dust to see what he is doing. Many cockroaches seek harborage and breed in the insulation in the walls of a refrigerator, stove, or range. These insects should come in contact with insecticide residues when they come out to forage for food and water. Insecticide sprays should also be applied behind window and door frames, under and behind furniture standing against the wall, on the undersides of tables and chairs, on closet and bookcase shelves, and in cracks and apertures around cabinets. Cockroaches are attracted to warmth, and may hide under and behind radiators and water heaters or in television sets and radios. Apply enough spray to wet surfaces thoroughly, but not enough to drip or run. If both sprays and dust are to be used, spray first and apply the dust after the spray dries, forcing the dust into cracks and voids.

When cockroaches are abundant, they are likely to be located in wall voids. In apartment houses, they may migrate from heavily infested to lightly infested or uninfested apartments via these voids. After blowing an insecticide dust into openings around utility pipes or any other openings into walls, particularly in kitchen and bathroom areas, such openings should be sealed to prevent further ingress or egress of the insects. Holes through which to dust wall voids can be drilled through plaster if necessary, and the holes can then be filled with matching plaster.

Holes can also be drilled at the top of the kick panel under a kitchen cabinet, or in the floor of the cabinet, in order to dust the subcabinet voids, unless the latter can be reached by removal of drawers or through existing cracks or apertures. After treatment, seal all cracks and crevices where cockroaches may hide. Even very narrow cracks may harbor the small nymphs. If the backsplash behind the sink and on the sink cabinet has parted from the wall, leaving a narrow aperture, this is a particularly favoured harborage for cockroaches and should be sealed.

Vapours and Gases in Enclosed Spaces

A piece of dichlorvos resin strip 1 sq in. -(2.5 cm) in area, attached to the underside,of the lid of a 24-gal (91-L) garbage can, results in complete mortality of all active stages of any of the 4 principal species of cockroaches in 24 hours under room-temperature conditions. A "ministrip" (2 x 2.5 x 0.25 in.; 5 x 6.5 x 0.7 cm) is now available that has an adhesive substance on one side so that it can be pressed against the bottom of a garbage-can lid. The ministrip, placed in closed kitchen sink cabinets in apartments infested with German cockroaches, resulted in 75 to 90 per cent, mortality during the first 7 to 12 weeks, decreasing to 50 per cent mortality at the end of 12 weeks. The results with the ministrip were not significantly different from those with a spray of 1.5 per cent malathion plus 0.5 per cent diazinon.

The efficacy of resin strips depends on the concentration of dichlorvos vapour they can generate in the area to be treated. In a normally ventilated room, the concentration does not reach a sufficiently high level for cockroach control, but the strips are effective against them in enclosed spaces with very poor ventilation or none.

Dry ice can be used as a safe and inexpensive means of generating a gas for exterminating cockroaches in food service carts, as used, for example, in hospitals. These carts have many areas where cockroaches can hide and be inaccessible to sprays or dusts. Tompkins and Cantwell (1973) developed a dosage and time schedule for killing not only active stages, but also the eggs. The cart was centreed on a 12 x 20-ft (3.7 x 6.1-m) sheet of black 6-mil plastic. Twenty pounds (9 kg) of dry ice was broken into pieces of appropriate size for placing at different levels in the cart. The plastic sheet was then drawn up and tied at the top. The volume of the enclosed cart was 71 cu ft (2 cu m). The level of carbon dioxide was monitored with a Gow-Mac Gas Analyser at 5 and 24 hours. At 78 °F (26 °C), the minimum concentration of carbon dioxide required for the 24-hour period of the test was 21 per cent in order to kill not only all the female German cockroaches with attached oyythecae that were used as the test insects, but also to prevent hatching of their eggs over the 30-day period of observation. This method of fumigation has the advantage of being safe and not requiring special training of personnel.

Boric Acid for Cockroach Control

Experiments made in our entomology laboratory at UCLA showed that the principal factor affecting the effectiveness of an insecticide against cockroaches was its repellency. The greater the repellency of the insecticide, the greater the chances that a certain percentage of cockroaches would learn to avoid treated areas before picking up lethal doses. Cockroach control is ordinarily accomplished by "spot treatment" of the dark and hidden places normally frequented by cockroaches during daylight hours. Some of the surviving insects in these areas may learn to avoid the treated areas and seek harborage elsewhere in the building. In apartment buildings, some cockroaches may leave treated apartments, and may invade untreated apartments via the wall voids or hallways.

Boric acid powder is the least repellent of the insecticides tested. In buildings treated with boric acid, cockroaches will crawl over the deposits repeatedly. Although they do not die as rapidly as when treated with more toxic insecticides, few if any should be seen after a week or 10 days. Boric acid is inorganic, and in an area in which it can remain dry and is not likely to be removed, it will continue to kill cockroaches indefinitely. Before using, it must be passed through a sieve or screen to break up lumps and facilitate its passage through the usual dust applicators.

Sodium fluoride is likewise an inorganic powder, and is much more toxic to insects and to humans than boric acid, but it is highly repellent, and has been far less effective in control of cockroaches, both in laboratory devices utilized to simulate typical harborage areas in buildings (choice boxes, mockup wall voids, mockup closets) and in experiments made in apartment buildings. As applied for cockroaches, boric acid is also effective for control of silverfish.

Boric acid acts in part as a contact insecticide, penetrating through,the insect's cuticle. In addition, cockroaches remove the dust from their antennae and legs while passing these appendages through their mouthparts. In this way, much dust is consumed, and boric acid, like sodium fluoride, acts partly as a stomach poison. The point to which boric acid, stained with Janus Green dye, generally penetrates into the alimentary canal of an adult German cockroach within 3 hours when the insect is placed on a deposit of 2 cc of boric acid in a petri dish. Repeated trials showed that the powder always filled the crop of a cockroach to capacity within this period, and although the insect might continue to preen for another 2 hours, its esophagus, pharynx, and mouth were packed with powder and no more could be ingested. The powder in the crop is prevented from moving farther into the alimentary canal, except in very small quantities, by the denticles of the armarium of the proventriculus.

In areas in which a dust would not ordinarily be applied, the liquid insecticides discussed earlier can be employed advantageously for a liquid - dust combination treatment. Experimenting with 0.5 per cent chlorpyrifos spray alone or in combination with 2 per cent propoxur bait, 2 per cent diazinon dust, boric acid, or Drioneýÿ (silica aerogel plus pyrethrins), found that, in apartments with bad sanitation, only where dusts were used was control of German cockroaches satisfactory 2 months after treatment. They recommended the use of "a combination of inorganic dusts and insecticidal sprays that can be applied in appropriate and somewhat mutually exclusive areas, so that the efficacy of both the formulations is fully realised." When boric acid alone is used as the residual insecticide, pyrethrin aerosols can be applied in areas of heaviest cockroach concentration to hasten control of the insects. Aerosol canisters, are suitable for this purpose, but have the disadvantage of not being refillable. A refillable aerosol generator with a flexible hose, such as the Hi-Fog, can be used just as effectively and also more economically in extensive treatments.

The cockroaches should then be swept up and destroyed as soon as possible after treatment, otherwise many may recover. Boric acid dust should then be applied in the usual way. The efficacy of the pyrethrin-boric acid treatment, as determined in extensive field experiments, was later confirmed in other field experiments.

Japanese entomologists found that 20 per cent boric acid, in starch-bait tablets or granules, was not repellent to German cockroaches. The addition of 0.5 per cent lindane or fenitrothion resulted in an increased initial rate of mortality, but because of their repellency, these insecticides also increased the period required for a 100 per cent kill. Methyl myristate at 0.02 to 0.2 per cent or maltose at 10 to 15 per cent concentration increased the amount of bait consumed. With a given amount of bait, small granules scattered widely were more effective than the larger tablets not so widely distributed. Another advantage would be that granules are less likely to be found and consumed by small children and pets.

Application of Boric Acid During Construction. As an inorganic dust, boric acid lends itself well to "insectproofing" at the time of construction. The dust is applied in enclosed spaces, such as attics, dropped ceilings, wall voids, voids under cabinets, and built-in appliances such as ranges and refrigerators. Half-inch (12-mm) holes can then be drilled into wall voids and subcabinet voids to facilitate the application of the dust.

The treated areas are not only eliminated as harborage and breeding places for cockroaches but, with the relatively nonrepellent boric acid dust, they act as traps to eliminate cockroaches that may develop in untreated areas of the building. A water-type fire-extinguisher is particularly effective for this kind of dust application. Although cockroaches can still gain entry and develop in the "living space" of a building, they will never become so abundant as they would have if boric acid had not been applied in enclosed and inaccessible areas of the building at the time of construction.

Toxicity of Boric Acid to Humans. Boric acid has long been used in medications, and although in recent years such use has been greatly diminished, there is much confusion as to its potential hazard when used as an insecticide. Its LD50 is relatively high, but it should be borne in mind that boric acid is used undiluted or only slightly diluted as an insecticide, whereas most insecticides are used only in greatly diluted form. Nevertheless, there appears to be a considerable margin of safety in the use of boric acid. Absorption of the material in unbroken skin is negligible. A standard teaspoonful of boric acid, screened to break up lumps, weighs 2.2 grams. Sublethal doses are rapidly excreted in the urine.

When 2 normal women were given approximately 2 grams of boric acid during the third, fourth, and fifth days of a 28-d. metabolism experiment, 93 to 94 per cent of the boron was recovered during the first week, demonstrating its rapid absorption and metabolism. Chronic intoxication has occurred in man after daily ingestion of 4 to 5 grams of boric acid for 3 to 4 weeks and 6 to 20 grams of borax for as long as several months. On the other hand, death has occurred in adults within 46 hours after oral ingestion of as little as 7.5 grams and within 46 hours after the oral ingestion of 15 grams. Although infants have survived doses of 4.5 to 9 grams of boric acid in several hospital accidents, death to newborn babies has resulted from accidental administration of 3 to 13 grams. These figures show that death from boric acid is likely to occur only from accidental ingestion of considerable quantities. As in the case of any insecticide, as well as many common household powders and liquids, containers should be conspicuously labeled and kept out of reach of small children.

Control in Sewer Systems

In the warmer areas of the United States, cockroaches inhabit the sewer systems of many cities, sometimes in enormous numbers. There is some danger of cockroaches moving from sewers to buildings, particularly if sewage backs up into manholes, forcing the insects to flee. The species most commonly involved is the American cockroach, although the oriental cockroach has occasionally been found in large numbers. On the other hand, the oriental cockroach is commonly found in the water-meter vaults in alleys and parkways, while the American cockroach is never found in them.

Good control of the American cockroach in sewer manholes was obtained by blowing Drioneÿÿ, a silica aerogel-pyrethrin formulation, into the manholes at the rate of 0.3 lb (135 g) per manhole. The dust was applied with gasoline powered centrifugal blower of 500 to 1,500 cfm (14 to 42 cu m per minute) capacity. Even with this small quantity, dust was seen rising from plumbing vents protruding from the roofs of houses on both sides of the street and from manholes up to a block away, indicating good penetration of all sewer-system laterals. The same insecticide and procedure were used to control cockroaches in Yakima, Washington. As in the treatments noted by Wagner, no live ccokroaches were found immediately after treatment, but 2 miles away the sewage-disposal system was flooded with the insects that had been washed from the treated sewer manholes and laterals. Periodic inspections showed little reinfestation over a 2-year period, but additional infested areas found during routine sewer-maintenance work were treated in the same manner, and this method continues to be used. Other cities in the Pacific Northwest have adopted the same treatment.

In Phoenix, Arizona, good results were obtained when carbaryl insecticide dust was blown into manholes, using a portable crop duster that gave the dust a 36,000-volt electrical charge as it left the blower spout.

Application of a 50 per cent chlordane emulsion in sewer-system manholes for, American cockroaches and in meter boxes for oriental cockroaches was also found to result in good control in San Diego, California. Even the spraying of only the upper 3 ft (1 m) of the manhole was successful in controlling American cockroaches in one experiment. The control campaign involved the treatment of 14,500 manholes at a cost of 22 cents each and 150,000 water-meter vaults at 4.5 cents each, including the cost of materials, labour, and equipment.

Chapter 9

Ants

Ants are among the most common pests of the home. Their nests are found almost everywhere - in lawns, under the house, in the walls of the house, in open ground, under stones or wood, in wood of decaying trees, or in other locations that afford protection. Ants are social insects of the order Hymenoptera, and live in colonies. They cannot live alone, but unlike other social insects, they are not dependent on a definite habitation, but can quickly transport their eggs, larvae, and pupae to a more favourable location when necessary. With the exception of the males, ants share with termites the advantage of a long life span and comparative freedom from natural enemies.

Ants enter buildings when seeking sweet or fatty substances in the kitchen, pantry, storeroom, or warehouse. They may also be pests of the lawn or garden because of their nesting habits or because they may destroy plants, seeds, fruits, and nuts. Ants may feed on the sugary excretion (honeydew) of certain plantsucking insects, such as aphids, psyllids, mealybugs, unarmored scales, cottonycushion scales, whiteflies, and treehoppers. In crawling. back and forth over the infested plant, the ants interfere with activities, such as oviposition, of the natural enemies of these pests. In addition, ants actually destroy adult parasites and the larvae of predators. They may also carry aphids and other honeydew-excreting insects to plants in order to establish colonies of them. Control of ants is a great aid in garden pest control.

How to Recognize Ants

The principal body regions of ants are distinctly defined by narrow constrictions. The elbowed antennae are composed of a *scape* (long basal segment) and a *funicle* or *funiculus*. At the end of the antenna, the last 1 to 4 segments may be enlarged to form the "antennal club." The abdomen is attached to the thorax via 1 or 2 slender segments that are parts of the abdomen, and form the *petiole* or *pedicel*. Each segment of the petiole has a bump or *node* on its dorsal surface. Some genera have 2 nodes, and other genera, higher in the evolutionary scale, have 1 node, so the number of nodes is one of the morphological features by which an ant can be identified. The remainder, and major portion, of the abdomen is called the *gaster*. It may consist of 7 or 8 segments, the male having 1 more than the female.

Ant Bites and Stings

Some species, such as the harvester ants (*Pogonomyrmex*) and fire ants (*Solenopsis*), inflict severe stings. Others merely bite, but they may curve their abdomens forward to inject a secretion into the wound and are, therefore, thought to be stinging (e.g., *Formica* and other Formicinae). The Formicinae are the only ants that secrete formic acid, emptying it into the wound. The venom of ants that sting does not contain formic acid, neither does the venom of wasps and bees; therefore, local application of alkalies results in no relief from pain when one is stung by such insects.

Ants that have stingers normally use their venom offensively only to kill prey, but if the nest is attacked, they may use the venom defensively to repel the intruders. In the present chapter, only the species are discussed that are pests mostly because they nest in the house or invade it from the outside, and became nuisances by getting into household foods.

Iridomyrmex Pruinosus (Roger)

This is a native species that ranges from Wisconsin to New York, south to New Mexico and Florida, and also occurs in

Colorado, Arizona, Utah, Nevada, California, Oregon, and Idaho. It is common and widely distributed in the southern United States. It infests houses, particularly in the Gulf Coast states, generally invading them from outside, but is also believed to be able to nest within a house.

Description. this species can be distinguished from the Argentine ant by the difference in the shape of its head, the presence of erect hairs on the thorax, and by having a more densely pubescent body. The body colour is variable. In the eastern states, this species is commonly uniform dark brown or black, whereas the body of the Argentine ant is usually light to dark brown. In the western states, the body colour is even more variable, some specimens being even lighter in colour than *Iridomyrmex humilis*. The workers are 1.8 to 2.5 mm long. The odor of crushed specimens of *I. pruinosus* is like "rotten coconut," similar to the odor of the odorous house ant (*Tapinoma sessile*), whereas a mass of crushed specimens of *I. humilis* has a stale and musty or greasy odor.

Habits. *Iridomyrmex pruinosus* prefers open spaces, such as fields, meadows, pastures, or entirely bare areas, but will also nest in open woods away from dense and prolonged shade. The nests may be constructed in exposed soil or under cover of stones or the bark of logs and stumps, and contain small to moderate-sized colonies. Like the Argentine ant, this species feeds on live or dead insects, is particularly fond of honeydew, and forms foraging trails.

Harvester Ants

These ants are predominantly black, and have conspicuous thoracic spines. The petiole has 2 nodes. Their habits are somewhat similar to those of *Pogonomyrmex*, except that they do not sting, their stinger being vestigial and nonfunctional. Homeowners object to their nests in yards and lawns. They are very active, and harvest large quantities of seeds.

Longspined Harvester Ant, Veromessor Andrei

This species is 4.5 to 7 mm long, with a reddish-brown head and thorax, and in some specimens the posterior part of the abdomen is black. The remainder of the body is lighter in colour - in southern California, generally reddish brown. It is limited almost entirely to California in its distribution. Wheeler and Wheeler (1973) point out that while *V. andrei* does not sting, its bite is quite tenacious and annoying.

Jetblack Harvester Ant

This species is 2.5 to 6 mm long, and as the common name indicates, is a beautiful "jet-black, patent-leather-coloured" ant (Mallis, 1941a). It inhabits the deserts of southern California, northwestern Arizona, and Baja California. The nests of *V. pergandei* are usually surrounded by a crater of excavated soil, whereas those of *V. andrei* usually are not. *Veromessor pergandei* makes a neat pile of chaff with seed husks in the form of a semicircle or ring around the nest, whereas *V. andrei* makes a more ragged pile or scatters the chaff about. *Veromessor pergandei* can tolerate the most extreme heat of the deserts, but *V. andrei* cannot.

Bigheaded Ants

These ants are so named because of the exceptionally large heads of the soldiers in relation to their bodies. They live in situations somewhat similar to those inhabited by fire ants, have similar habits, and are distributed throughout the warmer areas of the United States. As in the case of the fire ant, the petiole has 2 nodes. They resemble fire ants superficially, but can be distinguished from them by their 12-segmented antenna, 3-segmented antennal club, and the large head, particularly of the soldier. The natural foods of *Pheidole* spp. are generally live and dead insects, seeds, and honeydew, but in houses they will feed on meats, grease, peanut butter, fruit juice, and other foods.

Western Bigheaded Ant

This species may be found throughout the San Joaquin and Sacramento valleys of California, as well as in most of the state below 3,000 ft (900 m) elevation, and east to Texas. These ants are 3 to 6 mm long, uniformly light to dark reddish, with the soldiers sometimes darker and more brownish than the workers. The soldier's head is deeply furrowed in the frontal view. The body has abundant long hairs. *Pheidole hyatti* is common in arid and semiarid areas, and makes nests in the soil, often beneath stones and boards. It is usually nocturnal, and is omnivorous, with a decided tendency to be predaceous. The ants are sometimes found in and around houses; in kitchens, they are usually found around the sink. They are especially fond of greasy meats.

Acrobat Ants

Acrobat ants are characterized by the heartshaped abdomen, the upper surface of which is flattened and the lower surface curved. They may be black, brown, or yellow, and the petiole has 2 nodes. Their common name is derived from their habit of raising the abdomen over the head and thorax. They are slow moving ants that tend aphids, and are known to construct "tents" or "cowsheds" of earth and vegetable debris over the aphids from which they obtain the sugary excrement known as "honeydew". They are generally found under stones or in old stumps, and have sometimes been found to tunnel and nest in wood. They occasionally become pests by invading houses.

Lined Acrobat Ant

This ant ranges from southern Canada southward along the Rocky Mountains and eastward to Florida, and has been found at altitudes up to 5,500 ft (1,700 m). It varies in colour from light brown through dark brown to black. The scape (first segment) of the antenna extends beyond the posterior border of the head. The workers are 2.5 to 3.5 mm long. Medium-sized to large colonies may nest in exposed soil, under stones or other objects, or in logs, stumps, and dead trees. The ants may also nest in the woodwork of houses and infest household foods, where they show a slight preference for sweets and meats or other high protein foods. In nature, they feed on honeydew and dead or live insects, including winged termites. The workers bite aggressively, and give off a repulsive odor when alarmed.

Pharaoh Ant

This species is native to Africa, but has been spread by commerce to nearly every part of the world. The specific name, from which the common name was derived, resulted from Linnaeus' mistaken belief that the ant was one of the plagues of Egypt in the time of the Pharaohs, mentioned in Scripture, whereas the only plagues mentioned were lice, flies, and locusts.

Although the Pharaoh ant occurs throughout the United States and Canada, it is rarely found in California, but appears to be established in at least 2 locations in the state. It has never been identified as a household pest in California, but is important in some other parts of the country. It is especially common in hotels, large apartment houses, grocery stores, or other places where food is commercially handled. In residential areas, nests may be found outdoors in the lawn or garden and indoors in wall voids, subfloor areas, attics, cracks, crevices, and behind wainscoting, baseboards, plaster, or mantles, under hearthstones, between flooring, or in furniture. The ants tend to seek warm and inaccessible locations.

They breed throughout the year, and produce enormous numbers of individuals. They feed on dead and live insects, and often damage the dried specimens in collections. In the home, they appear to prefer grease, fats, and meats, but are practically omnivorous. In the early years of American agriculture, they caused great damage to crops. For example, they injured young corn plants by gnawing the blades and drinking the sap.

Monomorium pharaonis has become particularly well adapted to existence in close association with man by abandoning the nuptial flight and territorial boundaries. Colonies multiply by "budding." If groups of these ants in a large building are joined by odor trails, they may constitute a widely dispersed colony that can contain up to millions of workers and thousands of queens.

Description: The Pharaoh ant is one of the smallest ants, 1.5 to 2 mm long, and is light yellow to reddish in colour. The petiole has 2 nodes. It is often confused with the thief ant, *Solenopsis molesta*, but averages slightly larger, and can be distinguished by 3 segments in the antennal club, compared with 2 in the thief ant.

Life Cycle: Although the usual sexual forms are winged, nuptial flights have never been seen. Mating takes place in the nest, and the formation of a new colony is not normally accomplished by a single female after a nuptial flight, as with many other ant species. Instead, a large part of a colony will migrate to a new location, the workers carrying with them eggs and young stages. In this way, new nests can be developed throughout a building, and such nests have been found to contain from 400 to 4,800 ants. In subsequent investigations, they found that the various life stages of the workers required the following numbers of days: egg, 5 to 6; larvae, 22 to 24; prepupal, 2 to 3; and pupal, 9 to 12; thus, the life cycle required 38 to 45 days at room temperature. Sexual forms required about 4 days longer. The maximum period of adult life of the workers was 9 to 10 weeks, females lived as long as 39 weeks, and males generally not more than 3 weeks. The females were able to lay from 350 to 450 eggs.

Habits: Pharaoh ants crawl great distances from their nests, and do not always follow definite paths. Therefore, the nests may be difficult to find, increasing the difficulty of control. A shipowner spent $4,000 in trying to eradicate Pharaoh ants from his vessel. They have been stated to be the most persistent and difficult of all our house-infesting ants to control.

An interesting aspect of Pharaoh ant infestation is their tendency to nest in odd places, as between sheets of stationery, layers of linen, between clothing in trunks, in an electric iron, in a can of modeling clay, or, in piles of trash.

Many investigators have noted the tendency of the Pharaoh ant to be attracted to water, such as from dripping faucets, and one observer noted that their trails to the earth beneath the house did not necessarily indicate the presence of a nest, for the ants might be seeking moisture in the soil.

Little Black Ant

This small, jet-black species is 1.5 mm long. Its petiole has 2 nodes. It is found throughout the United States. In California, it occurs in the southern part of the state and in the San Francisco Bay area. These ants may nest beneath rocks, in rotting wood, in lawns, or in open areas, where the nests are indicated by very small craters of fine soil.

Their colonies are usually moderate to large, and contain many fertile females. Males and winged females may be seen from June to August. The workers are predaceous on other insects, tend honeydew-producing insects, and feed on honeydew as well as nectar and pollen. There have been reports of recently hatched bluebirds and kingfishers being killed by these ants. They may either invade a house from the outside, or establish their nests in the woodwork or masonry of the building. In the house, they feed on both sweets and greasy foods.

Thief Ant

Thief ants are found over most of the United States, and have been found at altitudes up to 5,500 ft (1,700 m). In nature, the nests of *S. molesta* and other tiny thief ants are usually located so that extremely narrow passages from the thief ant nests communicate with the nest passages of a larger ant species. The common name of the thief ant was derived from its habit of steadily pilfering the brood and other food from the nests of larger species. The stinger of the thief ant is rarely used. Although they are in the genus *Solenopsis*, thief ants are not considered to be "fire ants."

Description: The workers are very small, 1.3 to 1.8 mm long, among the smallest ants in the United States. Tle workers are monomorphic (1 form only). The petiole has 2 nodes. The body is largely smooth and shiny, and ranges from yellowish to light or dark brown. Thief ants superficially resemble Pharaoh ants, but are smaller, lighter in colour, and the antenna has a 2-segmented rather than a 3-segmented club. The antennal club is unusually large and elongate, about 1.33 times the length of the remainder of the funiculus. The scape of the antenna extends farther than between the eye and the posterior border of the head.

Biology: The outdoor nesting habits of the thief ant are similar to those of other *Solenopsis* species. There may be several hundred to a few thousand ants in a colony. In nature, they nest in exposed soil or under stones and other objects, and in rotting wood. Copulation takes place during nuptial flights, which have been observed from late July to early fall. A single fertile female is able to establish a nest.

Habits: Workers are predaceous, and feed on both live and dead insects. They also feed on planted or germinating grains, often causing considerable damage. They seek honeydew and tend honeydew excreting insects. Because they feed on dead rats and mice, they may be carriers of disease-producing organisms to food. They are known to attack young chicks in poultry houses. Workers have been seen carrying gravid segments of the poultry tapeworm, *Raillietina tetragona* (Molin), into their nest, suggesting that the ants may be intermediate hosts for the tapeworm.

When they are found in the home, thief ants usually have their nests in wall voids or elsewhere in the house rather than entering from outside. Their extremely small size enables them to enter containers not accessible to larger ants. They often infest pantries, cupboards, cabinets, and shelves, even though these have been kept closed and scrupulously clean. They are often found around sinks. However, they are so small that they are sometimes not seen, even when crawling on food.

Thief ants will eat practically anything, but appear to prefer foods with a high protein content. They have more of a tendency than the Pharaoh ant to feed on greases, but will eat sweets also.

Pavement Ant

This species occurs in most of the eastern half of the United States and in California, where it occurs in both the San Joaquin and Sacramento valleys. Nests occur under stones, along edges of curbing, and in cracks in the pavement. The nests may or may not be surrounded by craters. When nesting in houses, these ants are most likely to be found around or between the lower masonry walls of the foundation. In some areas of the eastern states, the pavement ant is considered to be the most important ant pest in houses, where it may be found throughout the year, but in greater numbers during the summer. It feeds on both greasy and sweet foods.

Description: This species is 3 to 4 mm long, and its body varies from light to very dark brown, with lighter legs and antennae. The head and thorax are furrowed with parallel lines, and the posterior dorsal portion of the thorax bears a pair of small spines. The petiole has 2 nodes. The entire body is covered with stiff hairs.

Velvety Tree Ant

The principal pest in the genus *Liometopum* is the California velvety tree ant, *L. occidentale*. There is only 1 other species in the genus, *L. apiculatum* Mayr, which ranges throughout approximately the same areas as *L. occidentale*, except that it does not occur in California.

Description: This species is 2.5 to 6 mm long, with a glistening, velvety-black abdomen, red thorax, and brownish-black head. Except for the 2 species of *Iridomyrmex*, which because of their economic importance and well-known biologies were used to introduce the ant section of this chapter, the preceding ant species all have petioles with 2 nodes.

Habits: Throughout California, this ant occurs in large colonies in the crotches or hollows of trees, in stumps, and in the ground beneath stones. The insects can be found crawling up trails, up and down oak, alder, elm, sycamore, cottonwood, and other trees, particularly along streams, foraging for insects and honeydew. Areas along streams and under trees also happen to be popular picnic grounds, and the velvety tree ant is therefore the bane of picnickers - biting people, crawling over their food, and producing a disagreeable odor when crushed. The removal of decayed portions of ornamental or shade trees in which the ants nest helps to eliminate these pests.

Pyramid Ant

This small ant is occasionally encountered at lower altitudes throughout California, where it nests in dry and vegetation-free areas. The earth excavated from its nest and gallery system is in a circular crater, usually about 2 to 4 in. (5 to 10 cm) or larger in diameter. It invades the home from the outside, feeds on a wide variety of household foods, but seems to prefer sweets. This species is widely distributed over most of the United States, but is more common in the southern half.

Description and Habits: The pyramid ant is 1.5 to 2.0 mm long, and has a reddish-black head and thorax and black abdomen. Its common name was suggested by the pyramidlike projection on the thorax. The petiole has 1 node. The workers travel in trails, and are fastmoving. They seek honeydew on garden plants, but are also carnivorous and predaceous.

Bicoloured Pyramid Ant, *Dorymyrmex bicolour* Wheeler

This species is similar to *D. pyramicus*, of which it was once considered to be a variety, but it is slightly larger and has a red head and thorax. It is found in the same geographical areas as the pyramid ant and in similar locations, but is of very little importance as a pest.

Odorous House Ant

The odorous house ant, so named because of its unpleasant odor when crushed, occurs throughout the United States, from sea level to 10,000 ft (about 3,000 m) elevation. It is second only to the Argentine ant as a household pest. However, in areas where the Argentine ant is also found, the odorous house ant tends to become less common or to disappear entirely. Nevertheless, it still frequently invades homes or apartments in southern California. The odorous house ant had a tendency to invade houses late in the year when the honeydew upon which it generally fed became less abundant. Also, rainfall and leaf fall, which reduce the numbers of aphids and other honeydew-excreting insects, may induce the ants to enter houses in search of food. It is most likely to invade houses during rainy weather. This ant does not possess a stinger.

Description: The odorous house ant workers are 2.4 to 3.25 mm long-slightly larger than the Argentine ant. The petiole is a single segment, with the node vestigial, flattened or inclined, and concealed from above by the base of the gaster. Like the Argentine ant, the odorous house ant travels in trails, and it also seeks the same kinds of food, preferring sweets. It can be distinguished from the Argentine ant by its darker colour, broader abdomen, by the fact that the petiole and its 1 vestigial node are hidden by the abdomen, and by its odor when crushed. Its pungent, "rotten-coconutlike" odor is detectable even when only a single individual is crushed.

Biology: The odorous house ant nests in many types of habitats - sandy beaches, pastures, woodlands, and bogs. It can also locate in buildings, and often nests in wall voids, particularly around hot-water pipes and heaters. Its nests may be under stones or logs, but also under loose bark, in plant cavities, insect galls, refuse piles, and bird or mammal nests. In the soil, the nests are shallow and indefinite in form. The number of ants in a colony may range from a few hundred to many thousands, and there may be many reproductive females. Mating usually takes place in the nest, but nuptial flights have been observed. Lone females have been observed to establish a colony, although colonies might also be formed when 1 or more fertile females left the parental colony accompanied by a number of workers. An average colony was found to have 345 eggs, 829 larvae, 533 pupae, 2,319 workers, and 1 to 6 dealated females.

Habits: As just mentioned, the workers travel in trails. They are very active, and have rapid movements. When alarmed, they dash about in an erratic manner, and the entire gaster is elevated. Odorous house ants are avid seekers of honeydew, and have been observed to transport honeydew-excreting insects such as aphids and mealybugs. They also feed on both live and dead insects. Although they will eat almost any type of household food, they prefer sweets.

Crazy Ant

This introduced species, possibly of Oriental or African origin, is now almost cosmopolitan, and is frequently intercepted in plant quarantine. It is common in the Gulf Coast region, especially in Florida, but is also well established in many towns and cities throughout the United States, including California. In the colder regions of the country, it occurs mainly in apartment buildings, hotels, and greenhouses, and can be a pest all year. Its common name axises from the worker's habit of running in a haphazard manner, as if lacking a sense of direction.

Description: The workers are 2.2 to 3.0 mm long. They have slender bodies, with very long antennae and legs. The antennae are 12-segmented, and do not have a club. The body is dark brown to black, with a "peculiar gray to violaceous luster or sheen," and has long, coarse hairs. The petiole has 1 node. The insect has no stinger.

Habits: Outdoors, the crazy ant nests in trash, cavities in trees, rotten wood, and in the soil under various objects in colonies of moderate to large size, but it can also nest in buildings. The workers run very rapidly, and can even jump, according to some observers. Their keen sense of smell enables them to find food quickly. They are almost omnivorous, feeding on live and dead insects, seeds, honeydew, fruits, plant exudates, and many household foods. In Puerto Rico, workers of *P. longicornis* have been reported to be predaceous on the larvae and adults of the house fly, flesh flies, and other Diptera. They have been observed to attack the larvae of the oriental rat flea under laboratory conditions, indicating that they may be important in reducing flea infestations under natural conditions. On the debit side, the workers are also known to gather small seeds of such crops as lettuce and tobacco from seedbeds, and their fondness for sweets forced the proprietor of a soda fountain in Florida to discontinue business.

False Honey Ants

False honey ants occur throughout the United States. The workers are 3 to 4 mm long, light to dark brown, very shiny in appearance, and have a somewhat triangular abdomen. They seldom invade the home. Their name probably comes from their habit of feeding on honey, honeydew, and nectar until their gasters are greatly extended, almost as much as those of the honeypot ants (*Melophorus*) of Australia and of *Myrmecocystus* of the United States Southwest. Honey ants also feed on the exudates of oak galls, live and dead insects, the fluids of dead earthworms, and the young, tender tissues of plants.

Small Honey Ant

This species ranges throughout much of the United States, is common in California, and often replaces the Argentine ant where the latter has been exterminated. When these ants are found in houses, they have usually come from outdoors, but it is likely that they often nest inside, for on many occasions winged females and males have been collected inside houses from January to March. This ant lacks a stinger and seldom bites, for it is not aggressive. A striking characteristic of *Prenolepis imparis* is that it is most active outdoors at temperatures between 35 and 55 °F (2 and 13 °C), and has been observed to emerge from its subterranean galleries at 6 °F (–14.4 °C).

Ants feed on such household foods as "sweet corn, corned beef, meats, cakes, breads, sugar, honey, syrup, watermelon, and fruits, but are especially fond of sweets." Outdoors, the ants can damage flower buds. H. M. Armitage observed *Prenolepis imparis* feeding extensively on tender, young growth and the calyxes and unopened petals of orange buds.

Description: Prenolepis imparis varies in colour from pale to dark brown or black, with the head and thorax often lighter in colour than the abdomen. The body is smooth and shiny. The thorax is small, slender, and almost cylindrical, and is divided into 2 parts by a strong constriction. The abdominal petiole is composed of a single segment (1 node). The antennae are 12-segmented, do not have a club, and the scape extends beyond the posterior border of the head by half the length of the scape.

The Nest: This ant normally constructs its nest in soil, preferably moist clay or loamy soil in shady woodlands, but seldom under stones or other objects. The nest has a central opening, surrounded by a crater of characteristic earthen pellets. A single gallery leads to the nest, which may be from 18 to 51 in. (45 to 130 cm) deep and have 6 to 40 lateral chambers leading from it. There are seldom more than a few thousand ants in a colony.

Life Cycle: In winter, an old colony contains a single reproductive female, winged males and females, and many workers. Usually in March to April, the overwintering males and females make their nuptial flight, although most of the mating appears to take place on the ground. The female then discards her wings and seeks a crevice in the ground in which to oviposit and start a new colony.

Species of Lasius

The Lasius species are relatively unimportant in California. The workers are monomorphic, active, emit a strong formic acid odor, but do not have a stinger. The antennae are 12-segmented and lack a club. The eyes are well developed, and are located closer to the posterior border of the head than to the base of the mandible. If ocelli (simple eyes) are present, they are small and indistinct. The abdominal petiole has only 1 segment, and the petiolar node is narrow in profile and either vertical or not strongly inclined.

Control of Ants

Recognition of species is important in ant control. The odorous house ant and thief ant may nest in houses, and their trails will lead to the nesting areas, which can then be treated. When they nest in large, inaccessible areas, such as attics, wall voids, and voids under cabinets, dusting with a chlordane or silica aerogel dust may be the most convenient and also the most effective way to obtain complete coverage and good control. Although the Argentine ant nests outdoors, it infests houses more frequently than any other ant species. The housewife may wish to treat for Argentine ants indoors in order to get rid of them quickly, but for effective and long-term control of this and other outdoor-nesting species that may invade the house, the main reliance should be placed on outdoor treatment.

The Argentine, odorous house, thief, jetblack harvester, field, and velvety tree ants travel in trails, and the trails help to trace them to their nests. Repeated spraying or dusting of trails can result in good ant control. Some ants are attracted to sweets only, and the thief ant is usually attracted to fats only.

The other species are attracted to both sweets and fats, except for the harvester ants (*Veromessor*) which eat mainly seeds. A knowledge of the food habits of ants is necessary in order to decide on what type of bait to use when control is to be accomplished by the use of poison baits.

For control of ants within a house, sprays of 2 to 3 per cent chlordane applied with a sprayer or paintbrush to baseboards and cracks where ants are seen, or a 50 per cent chlordane dust applied behind baseboards or in out-of-sight areas, are effective. It may be most convenient to apply an aerosol containing chlordane, holding the canister close to the sprayed surface so as to wet it visibly. If ant trails are found outside the house, a spray or dust may be applied next to the foundation in the general area of infestation. Aphid-tending species of ants, like the Argentine ant, may be controlled in yards and gardens by spraying or dusting the ground under the infested shrubs or trees. The shrubs or trees themselves should not be sprayed, for they may be injured and natural enemies of the pests may be destroyed. This should not only help to control the honeydew-producing insect pests that are favoured by the presence of ants, but should also decrease the ant population in the general area of the home and thereby decrease the chance of infestation in the house.

A "hose-attachment" sprayer is a great convenience, and is a time-saver if large areas need to be sprayed. Best results can be obtained with power sprayers, such as those used by commercial pest control operators, to spray the entire area around a house where nests may be located as well as tree trunks and subfloor areas. Some operators control ants by periodically spraying the entire foundation and the adjacent soil.

The homeowner can effectively use baits, such as 0.125 per cent Kepone or 0.075 per cent mirex, placing them in ant-nest openings, around foundations, and in other areas frequented by ants. It is important to use the proper percentage of toxic ingredient when placing ant baits. For example, dieldrin at 0.05 per cent concentration in syrup gave complete control of the odorous house ant, but at 0.5 and 1 per cent, the baits were repellent and the ants would not feed on them. Dieldrin happens to have amazing toxicity against the odorous house ant - it required a 67 times greater concentration of chlordane to obtain an equal mortality level. However, dieldrin is no longer registered for such household uses.

Poisoned syrups are still commercially available for ant control, and are effective against species attracted to sweets, such as the Argentine ant. The receptacles containing the syrup should be placed at intervals of 10 to 20 ft (3 to 6 m) around the foundation of the house. The syrups contain sodium arsenite, and it is not advisable to use them if children or pets are in the vicinity.

For the control of Pharaoh ants in hospitals, where they are considered to be potential carriers of disease germs, baiting with arsenicals to be the only fully effective procedure. One successful bait had the following ingredients: ground beef, 1,000 g; arsenic, 200 g; and honey, 80 g; mixed by hand, using rubber gloves. Ground liver was the most attractive bait, and did not require the addition of honey. However, it dried out too rapidly in warm rooms, and was therefore more effective when mixed with beef. The bait was placed in locked wooden bait stations. Ground meat was left 3 days before poison bait was used, to accustom the ants to the new food station. Baits were renewed as often as required to avoid a dry bait surface. If an ant trail was disturbed during the baiting period, the insects fled to their nest, and it took several days for them to re-establish a trail to the bait station.

Chapter 10

Flies

All winged flies have only 1 pair of wings, the hind pair being replaced by a pair of clubshaped organs (halteres) used for balancing. There are about 384 species of flies that are wingless or have only rudimentary wings. Further information on the morphology of adult as well as immature stages of flies, and a discussion of the 3 major taxonomic groups (suborders). There are about 200 such species, so only the more important are included here.

Except for the psychodids, in which all species lack mandibles and most species have no maxillary blades, although the maxillary palpi are retained. Among the nonbiting species, are the house fly, the blow flies, flesh flies, eye gnats, and *Drosophila* flies. The *labial labella* of these flies are large, soft, oval lobes that can be spread out to form a broad disk by which food can be collected and conveyed to the food canal of the *haustellum*. The Cyclorrhapha, despite having no specialized piercing mechanism, includes some species, the horn flies and stable flies, that can bite and pierce. The labium forms the piercing organ, and the labella, instead of being soft, spreading lobes, are a pair of small, hard plates at the tip of the *theca*, and bear eversible teeth internally.

In the United States, the house fly is still generally the most abundant species in houses, especially in the southern states, but other flies that resemble it, such as the little house fly (*Fannia canicularis*), can also be troublesome, mostly as "doorway pests." The face fly (*Musca autumnalis*) and the cluster fly (*Pollenia rudis*) are often more abundant than the house fly in houses in some areas, especially the midwestern states. Face flies are usually house pests only in the overwintering form. Resistance of house flies to various organochlorine and even to some organophosphorus compounds should result in their retaining considerable importance as household pests. One or two treatments per year are not likely to result in control, as they did in former years.

Methods on Fly Production and Fly Control

1. There is an increased number of poultry and other livestock;
2. The economics of shipping and marketing influence the location of livestock concentrations near metropolitan consumer centres;
3. Man's rapidly expanding suburban populations are bringing more people into closer proximity and daily contact with previously relatively isolated rural communities; and
4. Recent changes in agricultural technology economically favour large concentrations of animals.

The result is that in the fringes of large urban areas, we are developing high-density, low-area monoculture of beef and milk cattle, poultry, etc. These, like monocultures of plant crops, tend to create difficult pest problems. Monoculture is the cultivation of a single product over an extensive area to the exclusion of other possible uses of the land. This practice decreases the complexity of the ecosystem, and this is generally disadvantageous to the natural enemies of livestock or plant-crop pests.

Before the current practice of high concentration of livestock in limited areas, the animals scattered their feces in small piles over large ranges or pastures. This created a favourable ecosystem for minimal fly development. Poultry and hogs could consume various pest arthropods that developed in their feces, but this is no longer possible. In cattle feedlots, all important predators associated with pasture feces have been eliminated, along with almost all coprophagous arthropods that efficiently convert pasture feces into arthropod life (Anderson, 1966a). The importance of the latter factor is indicated by the estimates quoted by Anderson of the annual production of manure in the United States - nearly 2 billion tons per year by cattle alone. Sound management in manure disposal is essential for effective fly control programmes. A complete utilization of manures as fertilizers would appear to be a worthy goal toward which to work, and consistent with the current desire for "recycling" and efficient utilization of our resources.

Some flies can be recognized by superficial characters. Thus, the bluebottle flies of the genus *Calliphora* can be recognized by having a metallic blue abdomen but a dull thorax. The flies with dull-gray or brown to black bodies can be separated from those with the entire body shining metallic or black. Within these 2 major groups of flies, wing venation alone separates smaller groups, after which other structural characters are utilized for separating the species of each group.

Flies with the fourth vein angled are either house flies, *Musca domestica,* which have 4 black stripes on the thorax, or flesh flies, *Sarcophaga* spp., which have 3 stripes on the thorax. Flies with the fourth vein gently curved include the stable fly, *Stomoxys calcitrans* (piercing mouthparts) and the false stable fly, *Muscina stabulans* (sponging mouthparts). Flies with the fourth vein straight include the little house fly, *Fannia canicularis,* and the latrine fly, *F. scalaris.* Flies with the fourth vein sharply angled include the common greenbottle flies as well as the black blow fly, *Phormia regina.* The former have blackish anterior spiracles and relatively large dorsal thoracic bristles, while the latter has reddish anterior spiracles and smaller dorsal thoracic bristles. Within the genera *Phaenicia* and *Lucilia,* further separations can be made, and the CDC pictorial key includes the structural characters required.

House Fly

The house fly is cosmopolitan within certain latitudes, and is important not only because it annoys man and domestic animals, but also because its habit of crawling on garbage and excrement, as well as on humans and their food, makes it a potential carrier of various disease-provoking pathogens. Pathogens isolated from the house fly include those capable of causing diarrhea, dysentery, typhoid, cholera, poliomyelitis, intestinal worms, eye infections, yaws, anthrax, and tularemia. As with cockroaches, flies have been implicated as vectors of *Salmonella* bacteria, responsible for food poisoning and gastric infections. The house fly appears to be even more important as a potential vector of *Shigella* than of *Salmonella*.

It is not possible to incriminate the house fly as a transmitter of disease organisms with the same certainty with which, for example, *Anopheles* mosquitoes are identified beyond doubt in the transmission of malaria, or lice in the transmission of typhus. Flies are accused of spreading infections that can also be spread with the fingers or in contaminated food or water. They have been found to harbor over 100 different species of pathogenic organisms, and are charged with transmitting over 65 human and animal diseases, but only on circumstantial evidence. For example, outbreaks of diarrheal diseases are closely correlated with seasonal increases in fly abundance. Also, fly control has been closely correlated with a decline in the incidence of disease. An experiment made by the United States Public Health Service in Texas, that in towns sprayed with DDT, the rate of dysentery caused by *Shigella* bacteria declined, whereas in towns left unsprayed as controls, the disease rate remained unchanged. After 18 months, the sprayed towns were left unsprayed and the former control towns were sprayed. The relative incidence of dysentery in the 2 groups of towns was then reversed.

The scarcity of soundly documented cases of field transmission of disease by flies results in inadequate knowledge of how important they are in the natural transmission of the many pathogens commonly associated with them. Therefore, it is still principally the fact that they are annoying to have around domestic premises that motivates us to control them.

Description. The general size and appearance of the house fly are well known. It is 4 to 7.5 mm long, and may be distinguished from other house-frequenting flies by wing venation, the 4 dark stripes dorsally and lengthwise on the thorax, and the 2 velvety stripes, silver above and gold below, on its face. The eyes of the female are more widely separated than those of the male. With its sponging-sucking mouthparts, the house fly can ingest only liquid material, but some solid foods can be liquefied by means of regurgitated saliva, and thus ingested. All the aspects of house fly morphology, as well as biology and bionomics, were discussed in great detail by West.

Life Cycle. House flies lay their eggs in warm, moist materials that are suitable for larval growth, such as animal excrement, garbage, piles of lawn clippings, decaying vegetables and fruits, or soil contaminated with such material. Immature stages are shown in plate IV, 1, and figure 158. During warm weather, the eggs, which are laid singly but piled together in small masses, can hatch within a day. The cream-coloured larvae (maggots), distinguished by their being narrow at the head end, burrow into the food material in which they hatched, and may complete their 3 instars and pupate in a week or less.

As the pupation period approaches, the larvae seek the drier outer regions of their manure pile or other breeding medium, or may leave it to pupate a considerable distance away in the ground, in loose material, or under boards or stones. Under warm conditions, the pupal stage may last 4 to 6 days. Under the warmest natural conditions, a house fly may complete its life cycle in about 8 days, but this period is considerably longer at lower temperatures and when the larvae have had a less than optimum food supply. There may be as many as 10 or 12 generations in one summer. The adult flies generally live from 15 to 25 days, depending on temperature. In colder areas, the adults may overwinter. The potential increase in fly population in a single season is obviously enormous.

Sources of Infestation. In rural areas, piles of manure may be the principal sources of house flies, and in urban areas, piles of fermenting lawn clippings may be important sources of infestation. In garbage cans, blow flies generally breed more abundantly than house flies. The proximity of beef and dairy feedlots and poultry houses to residential areas has become an increasingly frequent cause of complaint and litigation.

People bothered with house flies are interested in knowing whether they live within the flight range of these insects from known sources of infestation, such as those just mentioned. An investigation was made of the flight range of flies that had been fed radioactive materials. Although some house flies were recovered as far as 20 miles from the release site, the bulk of the fly population did not move more than 1 or 2 miles.

Stable Fly

Another fly that superficially resembles the house fly but rarely occurs in houses is the stable fly. Unlike the house fly, it has a stiff proboscis projecting forward from the lower part of the head, enabling both males and females to inflict painful bites and suck out blood. Stable flies normally are serious pests of domestic animals, but on occasion they also attack humans. They are most likely to bite in the region of the ankle. Socks and stockings offer no protection.

Description. The stable fly is more robust than the house fly and has a broader abdomen. It can also be differentiated by the difference in wing venation (A and B of the figure), by a prominent, light-coloured area between the longitudinal bands in the middle of the thorax, and by the presence of a number of nearly round, dark spots on the upper surface of the abdomen. When the stable fly is at rest, its wings are held widely apart instead of projecting backward on top of the abdomen, as with the house fly. The stable fly rests with the head cocked upward and abdomen held against the resting surface, whereas the house fly often rests in a crouching position, with head down and body parallel to the surface.

Life Cycle. Stable flies breed in decaying hay, straw, or grain; piles of lawn clippings or rotting fruit; and various manures mixed with grain or bedding. They also breed in the manure of dairy cattle and in poultry droppings when they are "stacked." They are often found in large numbers on beaches, where they lay their eggs in piles of seaweed, and are sometimes called "beach flies." Unlike *Fannia* species, the immature stages of the stable fly somewhat resemble those of the house fly. In warm weather, adults may emerge in 20 to 25 days after the eggs are laid, but this period can be extended to several months at lower temperatures.

False Stable Fly

The false stable fly superficially resembles the house fly, but it is larger. Note also the difference in wing venation. It is closely related to the house fly, and has similar sponging mouthparts. The female lays her eggs in animal manure, human excreta, and decomposed vegetable matter, including lawn clippings. Twenty-five adult false stable flies were reared from the caked grass taken from beneath the blade guard of a rotary power mower. Several generations are produced during the summer months, and hibernation normally takes place in the pupal stage. The females often enter houses to oviposit, scattering from 140 to 200 eggs over various foods, particularly those that are slightly spoiled. Probably it is in this way that man occasionally consumes them and suffers intestinal myiasis.

Muscina assimilis (Falln) is darker than *M. stabulans,* and its legs are entirely black, while those of the latter species are in part reddish brown. In California, M. assimilis is a common species. Adults have been reared from partly decomposed watermelon and other vegetable matter. It is less likely to be found in homes than the closely related *Muscina stabulans.*

Black Garbage Fly

The genus *Ophyra* contains species that may be found near refuse dumps and poultry ranches and occasionally around cow and hog manures. The adults are not found in or around houses and other buildings, but congregate in trees and shrubs. The larvae are scavengers in decaying vegetation, animal matter, and excrement, but are also known to feed on the maggots of other flies. The black garbage fly is a small, black fly that originally invaded this country from Europe and Asia. One of its larvae may kill from 2 to 20 house fly larvae per day. Fortunately, the *Ophyra* adults rest at different sites from those frequented by adult house flies. This should make it possible to apply insecticides to sites upon which house flies and other nuisance flies rest, and to avoid those upon which *Ophyra* adults rest.

Little House Fly

Males of this cosmopolitan fly are often found in houses; the females are rarely found indoors. They are more uniformly distributed than the house flies throughout the average home. In one investigation, 65.6 per cent of house flies, but only 40.9 per cent of little house flies, were found in the kitchen. Males of the little house fly tend to fly about aimlessly in the middle of the room or some shaded area outside, often hovering or making jerky, zigzag flights.

The little house fly appears earlier in the spring than the house fly, and in regions of high temperature it tends to disappear during the summer. In many areas, this difference is possibly explained by the fact that the little house fly becomes active at a lower temperature, and also suffers heat paralysis at a lower temperature. This difference in temperature requirements may account for the fact that in subarctic regions, Fannia canicularis often replaces Musca domestica, and is, for example, the common house fly of Iceland. This and other species of *Fannia* are attracted to honeydew, and are often found swarming under aphis-infested vines or trees.

Description: The little house fly resembles the house fly, *Musca domestica,* but is smaller (3 to 5 mm long) and more slender. The wings in repose are parallel and partly folded over each other, whereas those of the house fly are held in a diverging position. The abdomen of the female *F. canicularis* is ovoid and gray, while the male's is tapering, with pale yellow, semitransparent patches.

Life Cycle: The little house fly breeds in animal excrement or decaying vegetable material. Floatlike appendages on the eggs enable them to float on top of semiliquid material. The larva has spiny processes along the sides and top of the body. These have fine, whiplike hairs that are said to propel it through semiliquid media. The larvae change from translucent white to light brown as they mature.

Like those of the house fly, *F. canicularis* larvae move to a dryer environment near the surface of their food medium before pupating. The pupae resemble the third-instar larvae, but are darker and somewhat shorter. At 80 °F (27 °C) and 65 per cent relative humidity, the approximate durations of the immature stages were recorded as follows: egg, 1.5 to 2 days; larva, 8 to 10 days; pupa, 9 to 10 days. The period from egg to adult was 18.5 to 22 days, and from egg to egg, 22 to 27 days. About 50 per cent of the male fly population died within 14 days, and 50 per cent of the females died within 24 days. In another experiment, the life cycle took 24 to 29 days.

Latrine Fly

The latrine fly resembles the little house fly in appearance and life history, but its thorax is bluish black, contrasted to the thorax of *Fannia canicularis,* which is blackish gray. Also, a grounded protuberance of the median point of the middle pair of legs can be used to distinguish *F. scalaris.* It prefers human to animal excrement, and has been known to cause myiasis in man.

Coastal Fly

This species, of South American origin, is widely distributed throughout the United States. It is a little over half the size of *Fannia canicularis,* and is shiny black in colour. It sometimes occurs in large numbers, but is much less important than the little house fly as a pest.

Fannia Benjamini Malloch

This species is known from various areas of the West, particularly California. The adult is 3.5 to 4.5 mm long, with a brown-pollinose, unstriped thorax and a dark abdomen without a trace of the yellow translucence found in some species, such as the male of *Fannia canicularis*. The antennae and palpi are yellow, as are also the legs, with the exception of the black tarsi. The female is much lighter-coloured than the male.

The expansion of city suburbs into the foothills, as well as the increasing use of such areas for recreational purposes, is bringing *F. benjamini* into sharper focus as a pest. It is common in California along streams, particularly in groves of oak trees. It is a very annoying pest, and persists in its attempts to enter the ears, nostrils, and eyes, apparently attracted by perspiration and mucus. This species and *F. canicularis* have been implicated as vectors of a nematode eye worm, *Thelazia californiensis* Price. The known hosts of this nematode, besides man, are dogs, cats, horses, sheep, deer, coyotes, and bears.

Cluster Fly

Cluster flies derive their common name from their habit of entering a house in the fall and gathering together in a cluster that may emit a sickly sweetish odor. They tend to gather and hibernate in secluded areas, often in wall voids, and hibernate in them. Cluster flies and face flies (*Musca autumnalis*) often become trapped in large numbers in wall voids, attracting larder beetles (*Dermestes lardarius*). The beetles feed on dead flies and then migrate to other parts of the house, causing the occupants to believe they have an infestation of carpet beetles.

In the spring, cluster flies again become pests as they migrate from wall voids to the living space, via window pulleys, around baseboards, and through other small openings, and then crawl sluggishly over the walls. They are usually found in windows and other out-of-the-way places, and do not fly about noisily like other house-entering calliphorids. In this respect, they are not such nuisances.

Description. Cluster flies closely resemble house flies, but are larger, more robust, and slower in their movements. They are about 8 mm long, dark gray, with checkered black and silvery abdomens, and with many golden hairs on the thoraxes of newly emerged specimens which may be lost in older ones. The stripes on the thorax are not so prominent as those of the house fly. The wings overlap when at rest. *Pollenia rudis* can be distinguished from all other synanthropic North American calliphorids by its abdomen being nonmetallic in appearance and being covered by a pollenlike dust. A striking characteristic of the biology of cluster flies is that the larvae are parasitic on earthworms. Eggs are deposited in cracks in the soil, hatch in 3 days, and the larvae enter earthworms at almost any point on the body wall. The total developmental period of the fly varies from 27 to 39 days. Its breeding habits result in the cluster fly being less of a hazard than other calliphorids as a mechanical carrier of disease organisms.

Control of Cluster Flies

Lawns have been treated with 2.5 per cent heptachlor granules at 3 lb per 1,000 sq ft (1.36 kg per 93 sq m), or with a spray of heptachlor 25 per cent emulsifiable concentrate at 0.25 pt in 3 gal (0.12 L in 11 L) of water per 1,000 sq ft to kill earthworms, and thus prevent the development of cluster flies. The lawn should then be sprinkled to force the toxicant into the soil.

Blow Flies

The blow flies (greenbottle and bluebottle flies) are large scavenger flies, with a metallic blue or green sheen. They may become abundant in and around homes, particularly if they are located near such places as slaughterhouses, meat-processing plants, and garbage dumps. However, it should be borne in mind that some of these insects are stronger fliers than the house fly, and at least a small percentage may travel great distances from the point of dispersal. For example, in one experiment, 13 per cent of the black blow flies, *Phormia regina*, were trapped between 4 and 28 miles (6.4 and 46 km) from the point of origin, and more than 8 per cent at or beyond 11 miles (18 km) compared with only 3 per cent of house flies trapped at or beyond 4 miles.

Blow flies are most active during warm, sunny weather. When it is cool and cloudy, they tend to retreat to resting sites, usually on vegetation adjacent to areas where they oviposit. *Phormia regina* tends to disperse more widely than the other species, and in warm weather it often rests on sunlit walls.

Some blow flies, such as *Phaenicia sericata* and *Eucalliphora lilaea,* freely enter houses, and because of their breeding habits and the fact that they feed and oviposit on food, mechanical transmission of disease becomes a distinct sanitary consideration. Meat or fish must be protected from oviposition by blow flies, not only for aesthetic and sanitary reasons, but also because the eggs and larvae of blow flies and some other species, when swallowed, may remain alive for some time, causing acute gastric or enteric disorders.

Blow flies have been known to breed in the carcasses of dead birds or rodents in attics or wall voids of houses. These may be sources, not only of adult flies, but even of the larvae, which are sometimes found crawling about in the living space of the house. Because of their large size, buzzing sound, and tireless flying about throughout the house, blow flies are generally more annoying than house flies. They tend to fly to windows, for like all flies, except perhaps the males of *Fannia,* they are attracted to bright sunlight.

Although blow flies usually lay their eggs on meat or dead animals, when meat is not present they may oviposit on animal excrement or decaying vegetable matter. One female may lay as many as 600 eggs. Larvae first feed on the surface and then burrow down to less-decayed material. The full-grown larvae burrow into the ground to pupate, usually in the top 5 cm of soil. Tlere are several generations a year. The winter is normally spent in soil by full-grown hibernating larvae.

Some Common Species

Some common examples of bluebottle flies are *Calliphora vicina* and *Phaenicia sericata* Meigen are about 10 mm long, less bristly than *Calliphora,* and metallic blue-green. These and many other species occur throughout the United States. Large numbers of maggots of *P. sericata* dropped onto the bed of a resident of North Hollywood, California, falling from a light fixture in the ceiling over a period of several days. These had bred in the carcass of a dead rat.

The black blow fly, *Phormia regina* (Meigen), is slightly more robust than the greenbottle flies and darker in colour. The thorax is almost black, with a metallic bluish-green luster, and the abdomen is darker, being metallic green. *Protophormia terraenovae* is even darker, and has a purplish cast. The colours of some common blow or bot flies, as well as certain other synanthropic flies, are well displayed in 15 colour plates.

Flesh Flies

The flesh flies are not so likely to enter houses as the related blow flies, but many species, including some carrion and excrement feeders, are considered to be synanthropes (domestic flies). The sarcophagids are medium-sized to relatively large flies. They generally range from about 10 to 14 mm in length, and are robust and dull grayish black, with 3 dark stripes on the thorax.

The more familiar species are characterized by a checkerboard pattern of spots that change from black to gray and back with changes in the incidence of light. Sarcophagids are larviparous, i.e., the eggs hatch within the female's body and she deposits the larvae on the food they will be eating. Both the common and family names are derived from the fact that the larvae of most species develop in decayed flesh and spoiling meat, although some kinds can also develop in manure or other animal droppings, and in food residue at the bottoms of garbage cans. The life cycle ranges from 8 to 21 days, and adults may be present from spring through late fall.

A representative flesh fly is *Sarcophaga haemorrhoidalis* (Falln). It is nearly cosmopolitan, and in North America it occurs from Oregon to Quebec and south to California and Florida. The adults are 10 to 14 mm long, gray, with the terminalia of the male red. The eyes are reddish brown. The larvae feed on dead insects, carrion, and excrement. Another common and widely distributed flesh fly is *Sarcophaga bullata* Parker. It ranges from British Columbia to Quebec and south to California and Florida.

Drone Flies

The drone flies are robust, yellowish flies that mimic bees in appearance. They do not cause serious annoyance to man or animals, and rarely enter homes or alight on food. They are often seen around flower beds, the males hovering in the air, almost motionless, but darting swiftly aside when disturbed. Most syrphids are important predators of aphids, but there are also saprophagous genera that are scavengers in semiliquid, rotting vegetable and animal material or, in the case of *Eristalis*, in putrid or stagnant water or in moist excrement. The larvae are called "rat-tailed maggots" because of the elongate breathing tube that allows them to obtain air even when the remainder of the body is buried in liquid material.

Control of House Flies and Similar Species

The major contribution to fly control is garbage pickup and disposal. In Los Angeles, for example, a city ordinance requires that garbage be wrapped or placed in bags and placed in covered containers to await pickup by the City Sanitation Division once a week, along with prunings, litter, lawn clippings, or other organic materials that are potential sources of fly infestation. The trash and garbage are dumped in areas which are continually being covered with earth by means of giant bulldozers.

This is a method of trash and garbage disposal that may be particularly suitable when an earth fill is desired by the community in a particular location, as for filling a ravine or canyon. Incineration is of course a suitable alternative, except where it may contribute substantially to air pollution. City sanitation divisions are relieved of much of the burden of garbage removal by the wide use of automatic garbage-disposal units in the home, for much of the organic waste material thus becomes part of the city sewage.

Supplementing the community garbage-removal service, decaying organic material and animal excrement on residential property should be deeply buried or otherwise disposed of by the homeowner, for house flies, as well as stable flies, flesh flies, bottle flies, blow flies, little house flies, and false stable flies, all breed in this type of material.

When urban areas are surrounded by areas of high concentration of poultry and livestock, freedom from flies in the city may depend heavily on manure management, water systems, general farm sanitation, and the wise use of insecticides in an "integrated control" programme in such areas.

Exclusion at Doorways

The usual type of window or fly screen (18-mesh) on windows, screen doors, and other openings to buildings is a great help in keeping ingress of flies to a minimum. However, screen doors that must be opened many times during the day cannot be depended upon to completely exclude insects, and may be unsatisfactory where human traffic is heavy or where little, if any, insect entry can be tolerated. In such situations, it is possible to exclude flying insects by means of a downward air current at the doorway. An air speed of 1,500 ft (450 m) per minute, approximately 17 mi (27 km) per hour, when the ratio of the distance through the doorway to its height was 0.37, was found to offer a satisfactory combination of efficiency with economy and acceptability. The installation in the doorway of a removable floor grid or grating that can be readily treated with a long-term residual insecticide or a low-friction finish has been suggested for the exclusion of crawling insects.

Sprays

Two types of spray are generally used for fly control within buildings: space sprays and residual sprays.

Space Sprays. These are applied in order to produce a fine mist that permeates all the air space within a building, so that the insecticide contacts the insects as they are flying about or at rest. The insecticide may or may not break down rapidly; the objective is to contact all the insects in a room and kill them rapidly. A mist can be produced by applying the insecticide in a petroleum base, such as deodorized kerosene (base oil), with an ordinary hand atomizer or any other kind of atomizer. The same objective is accomplished by using an aerosol canister or fogging machine, which generates even smaller droplets, requiring longer for the droplets to settle out of the air. It is desirable to remove pets and ornamental plants and to keep doors and windows closed after treatment for as long as it is convenient. Cover foods and cooking utensils.

Dichlorvos (DDVP) is commonly used for fogging such buildings as food-processing plants, industrial plants, warehouses, and theaters, using 1 gal of a solution containing 10 per cent of toxicant to 64,000 cu ft of space (about 4 L to 1,813 cu m). Manufacturing buildings should not be in operation, and should be vacated and not reoccupied until they have been well ventilated. Food should be removed from the buildings, and food-handling equipment should be covered with draping paper before fogging - or it should be washed *after* fogging. A suitable mask or respirator should be used by persons applying the fog. Dichlorvos is also applied as a 0.5 per cent spray with a low-pressure sprayer to localized areas that may be infested by flies and other insects that are attracted to food, such as cockroaches and ants. In another procedure, it may be added to a longer-lasting residual spray to obtain rapid knockdown and fuming action in cracks, crevices, and other confined areas, combined with prolonged residual efficacy.

Synergized pyrethrins or allethrin at 0.1 per cent concentration have been used effectively as space sprays, but the newly developed synthetic pyrethroid, resmethrin, is as much as 20 times more toxic when used at the same concentration, and no synergist is required. The knockdown rate is slow, but whereas with pyrethrins the flies that are knocked down may recover, with resmethrin they do not. Resmethrin also differs from pyrethrins in that its residues may be effective for several weeks on some types of substrate.

Residual Sprays. These sprays are applied to surfaces upon which flies generally rest. The insecticide is expected to kill flies crawling about on the deposit long after the spray is applied. Great impetus was given to this method of fly control by the original spectacular success of DDT. This insecticide, which came into general use after World War II, was not only highly toxic to house flies and most other household pests, but also had a long residual life.

Against house flies, the effectiveness of a spray residue ranged from 8 months on unsized cement surfaces to 18 months on smooth surfaces, such as glass, wallpaper, or finished wood. Within the home, fly control could be obtained for long periods by spraying surfaces on which flies commonly rested, such as window screens. Residual wall and ceiling sprays in barns resulted in about 95 per cent reduction in the fly popiiiation. Flies entering during the day were killed overnight.

Within a few years, house flies developed resistance to DDT and other persistent organochlorine insecticides. Later, the little house fly (*Fannia canicularis*) was shown to be resistant also. Both species, taken from 3 poultry ranches in southern California, were also shown to be resistant to a number of organophosphorus compounds, but to a smaller degree than to organochlorine compounds. Still, *Musca domestica* was found to be over 909 times resistant to coumaphos, 182 times to Ciodrin, 102 times to malathion, and 66.5 times to diazinon. *Fannia canicularis* was found to be over 174 times resistant to coumaphos, 81.9 times to Ciodrin, 59.1 times to malathion, and 6.8 times to diazinon. The organophosphorus compounds and carbamates do not have the long residual life inherent in most organochlorine compounds, and resistance is compounding the problem. Diazinon, fenthion, dichlorvos, dimethoate, malathion, naled, and resmethrin are recommended as outdoor space sprays. Instructions on labels should be carefully followed when using any of these insecticides.

In California, resistance has not yet developed with Gardona, except on a low level in one area. In Denmark, however, where its use has been more extensive, high resistance is reported. The period required for resistance to appear against a new insecticide is in proportion to the frequency of its.use. This can be delayed by using methods of suppressing the fly population that do not require insecticides, thus reducing the frequency of their application.

For the control of cluster flies, spraying the outside of a house in the fall with 2 per cent malathion or 1 per cent naled or ronnel has been recommended. Particular attention must be paid to the upper third of the wall, the overhang, and the eaves, especially on the south and west sides. Residual sprays, such as 2 per cent chlordane, have been applied to the surfaces of attics, basements, closets, storerooms, and other areas where cluster flies and face flies are seen, usually with a 2- or 3-gal (7.5- or 11-L) compressed-air sprayer. To avoid staining wallpaper and furnishings, such sprays should not be applied in bedrooms or in papered or well-furnished rooms. Poisonous sprays should also be avoided in food-preparation areas.

A 10 per cent chlordane dust is highly recommended for treatment of places that are difficult to reach with sprays. A 0.5 per cent solution of dichlorvos may be sprayed into window-pulley openings or wall voids.

Community Spraying. A community-wide effort in fly control can include spraying. The little house fly, *Fannia canicularis,* was effectively controlled in one town near a large poultry farm by blowing a mist consisting of 8 gal of DDT 25E and 0.5 gal of malathion 57E per 100 gal of water from roads and driveways over lawns, shrubbery, and onto buildings, at the rate of 1 to 10 gal per acre, depending on vegetative cover. One application was made in the morning on July and another on September. Although such results can also be obtained with currently registered insecticides if the target species is not resistant, it must be borne in mind that beneficial insect species are also eliminated. In the foregoing example, only a nonvector species was involved, for which most complaints resulted from the activities of the hovering males. In the current climate of ecological awareness, a similar control programme would probably not be undertaken. Such community-wide spraying programmes should be considered only under emergency disaster conditions. Spraying should not be considered as a substitute for a basic sanitary solution at the sources of infestation.

Insecticide recommendations change as flies become resistant to insecticides. Generally, pyrethrum, with a suitable synergist such as piperonyl butoxide to increase its potency, is added to the currently recommended insecticides, and the combination provides good fly knockdown and kill. With the continually changing statuses of pesticides, an appropriate state official should be consulted for advice as to which compounds are currently registered for the projected treatment.

Other Treatments

A property owner with a fly problem may not be responsible for or have any control over the sources of infestation, but may wish to know what he can do to eliminate the flies on his own property. No great detail on methods and materials will be given here because these are continually changing. A few examples are given, mainly to indicate the types of treatment available rather than specific insecticides. Specific recommendations are revised almost yearly, and are available from university extension personnel or federal and state agricultural extension literature.

The current tendency is to resume the use of space sprays for fly control, or to use baits and "vapour strips." Emulsifiable concentrate gives 4- to 6-week residual protection during fly season. Treatments should begin in May or June at the rate of 2 gal per 1,000 sq ft (7.5 L per 93 sq m) or to the point of runoff on ceilings, walls, support posts, and outside around closed doors and windows. Rachesky also recommended the use of dimetilan fly bands in garages, under breezeways, in covered porches, and on loading docks. The bands should be wiped occasionally with a damp cloth to remove dust. They are effective for more than a year. No vapours are given off by dimetilan bands; the fly must contact the band.

Dry baits of dichlorvos or dipterex are often very effective in substantially decreasing the fly population, particularly when scattered around garbage cans and outside of doorways. Inside the home, they may be scattered on floors and near windows, according to directions on the labels. The baits should be widely dispersed, to increase the probability that flies will find them and to decrease the hazard to children and pets. Stable flies and most blow flies are not controlled by these baits, but cockroaches, snails, and slugs are susceptible if they happen to be in treated areas. Liquid baits, in which molasses, syrup, or sugar are usually the attractants may be applied as sprays, using 1 gal per 1,000 sq ft, when added to any organophosphorus or carbamate insecticide currently recommended.

Baits tend to have a spectacular but brief effect. A permanent liquid-bait station has been suggested that is merely a chicken-watering unit with a cellulose sponge in the trough to prevent clogging by dead flies. Dichlorvos or trichlorfon at 0.1 per cent concentration in a sugar/water solution has been used successfully. A constant source of toxicant can be provided by inserting a dichlorvos resin strip in the reservoir jar.

Little house flies (*Fannia canicularis*) seemed to prefer beer cans to regular garbage. This led to the suggestion that insecticide be placed in about 2 in. (5 cm) of beer in open containers for fly control. This poisoned bait method was found to be effective. Obviously, the poisoned beer should be placed in locations where it cannot be reached by children.

Dichlorvos resin strips suspended from suitable room fixtures are effective against small, flying insects and are convenient to use. The strip is a solid polyvinyl chloride resin formulation, impregnated with 20 per cent dichlorvos (DDVP) and related products. It is 2.5 in. wide, 0.25 in. thick, and 10 in. long. In the period of its activ,ity, the vapour issuing from the strip is highly toxic to insects, but it is rapidly hydrolyzed by moisture. The strips should not be used in kitchens, restaurants, or other places where food is prepared or served, where infants or aged people are confined, or near fish tanks. One strip is sufficient for 1,000 cu ft (28 cu m) of space, and remains effective for weeks. A smaller strip of the same material has been made available.

This "ministrip" is only 2 in. (5 cm) long, and is sufficient for a volume of 100 cu ft (2.8 cu m) for the control of such insects as cockroaches, silverfish and clothes moths in enclosed spaces. The strips may also be placed in a closed garbage can, attached underneath the lid, for fly control. An additional procedure is indicated when treating for flies, such as cluster flies, face flies, and blow flies, that spend the greater part of the winter in attics, wall voids, and similar enclosed spaces. Insecticide formulations, such as aerosols, silica aerogels, or the currently available combinations of silica aerogel and pyrethrum that can be very thoroughly dispersed, may be used in such areas.

Dichlorvos resin strips may be suspended in attics, closets, and storerooms, provided they have very little air circulation. If the flies work their way into occupied rooms, a space spray or aerosol containing pyrethrum is suggested, repeating as necessary. Sometimes flies in a house will escape if windows are open and the screens removed.

Exposure of house flies for 3 days to baited chemosterilants, with light corn syrup as the attractant, resulted in a high degree of sterility with hempa, metepa, and tepa within acceptable limits of adult mortality. With 1 per cent concentration, 1,00 per cent sterility of males was induced after 72 hours of exposure. The chemosterilants remained active for at least 20 weeks. The tests were sufficiently encouraging to justify further tests in local house fly control with baited chemosterilants alone, and with an integrated programme of baited chemosterilants with an insecticide.

Control of Flies on Poultry Ranches

Poultry production is a major nation-wide industry. Flies breeding in poultry manure can be a serious problem when the poultry ranch is situated rather close to urban areas. The most important house pests among these flies are *Musca domestica* and *Fannia canicularis,* that develop in manure, and *Phaenicia* spp., that thrive on fowl carcasses and broken-egg wastes. All can fly considerable distances - up to 20 miles and may be attracted to food and shelter in homes and backyards.

As already stated, because modern poultry production is a monoculture, natural enemies of poultry pests are at a disadvantage, just as are the natural enemies of plant pests in agricultural monocultures. In plant monoculture, changes in cultural practices and modification of the pest control programme in such a way as to avoid the usual need for continually increased dosages of pesticides seem to have been first advocated in 1940 by the Swiss entomologist Schneider, in an extensive monoculture of gambier in Malaya, and are now generally referred to as "integrated control."

This term appears to have been first used, who stated: "If we are to escape the ever-tightening spiral of more complex problems and ever-increasing costs of control, then we must integrate chemical control with the natural factors influencing populations." While admitting that many of the strands of the ecosystem of which man is a part remain to be unraveled, they believed: "We are now in a position to take more intelligent steps toward an integrated control programme which will utilize all the resources of ecology and give us the most permanent, satisfactory, and economical insect control that is possible."

As in the control of plant pests, the current trend in combating pests of poultry and livestock is "integrated control." The number and variety of predators and parasites of filth flies is impressive. The predators include many species of beetles, bugs, flies, ants, earwigs, and mites, and there are some effective hymenopterous parasites. The species that are natural enemies of the flies in accumulated excrement or manure piles are quite different from the ones that are natural enemies of the flies in the undisturbed feces of livestock in pastures. The 2 groups of natural enemies are discussed separately in an extensive review.

The importance of developing manure removal and pest control practices that result in minimal reduction of the influence of these natural enemies of the pest fly population is obvious. Sound integrated control programmes must be preceded by studies on the ecology and behaviour of the pest species and the complexity of the ecosystem of which they are a part. An important element of fly behaviour studies is the investigation of resting habits during the day and night, indoors and outdoors, in urban and in rural areas, and when buildings are screened or not screened. A study in poultry ranches in northern California, in which a backpack vacuum-sampling machine and sticky fly tapes were used, resulted in much information on which species of "domestic" or "filth" flies were involved, which species of natural enemies, and also much useful information on their habits, particularly their resting habits.

In striking contrast to the widespread dispersion of the pest flies during the day, at night all fly species except *Phaenicia sericatu* that remained outdoors rested predominantly on branches of trees and shrubs, and those inside the poultry houses rested predominantly in the general ceiling area. The flies remained at their overnight resting places for 12 to 16 hours, depending on temperature. Sticky tape catches revealed that 85 per cent of resting *Fannia canicularis* and *Musca domestica* were caught in houses and 91 per cent of the beneficial predator *Ophyra leucostoma* and 95 per cent of the ichneumonid parasites were caught in trees and shrubs. The difference in the nocturnal aggregation sites of pest flies and certain of their natural enemies "seems to offer one potential approach to an integrated fly control programme on poultry ranches".

The University of California Agricultural Extension Service has published a circular in which integrated control procedures on poultry ranches are recommended. A concrete base beneath caged laying hens to catch droppings prevents house flies from pupating in the soil. Natural enemies can then destroy a greater percentage of immature flies. During the season when flies are a problem, only the top portion of manure should be removed, and a 6- to 8-in. (15- to 20-cm) pad should remain on the concrete base, or all manure may be removed on an alternate row basis. In either case, natural enemies from the manure that remains can quickly invade adjacent fresh droppings. If all manure is removed at one time, a period of 6 to 12 months is required to reestablish the natural enemies.

If insecticides are required, they should be applied only to overhead and peripheral portions of the poultry houses. If droppings are sprayed, nearly all the natural enemies that normally infest droppings will be killed. These natural enemies can destroy more than 95 per centof the fly population. Residual insecticides directed against adult flies and applied to the ceilings and peripheral parts of the poultry houses do not interfere with key natural enemies. Great care should be taken to avoid contamination of feed and water. Diazinon, dieldrin, lindane, and trichlorfon are no longer labeled for use in poultry houses.

In an attempt to control flies on poultry ranches without long-range residue buildup and adverse ecological effects, the relatively new organophosphorus insecticide Rabon's, in an encapsulated formulation mixed with the feed, has been fed to poultry, and this passes through the bird without being assimilated to any appreciable extent. It causes complete mortality of fly larvae in the manure, while leaving no residues in body tissues of the poultry or in the eggs (yolks). Another organophosphorus insecticide, coumaphos, mixed with cow feed, showed promise in the control of house fly larvae in the manure, and the experiments have been continued.

In one investigation, a dichlorvos formulation successfully controlled fly larvae and adults for about 7 weeks with 3 applications of 2 g of actual toxicant per square meter of breeding area in Florida poultry houses containing caged layers. The investigators emphasized that concurrent treatment of 2 life stages of the flies might speed up the development of resistance to the insecticide, and they believed that an alternative chemical that would be equally effective against both stages was needed. However, advocates of integrated control do not believe that insecticides should be used in this way on poultry ranches, not only because of the possibility of rapid development of resistance, but because the treatment is nonselective and kills all beneficial species. They suggest that its use should possibly be restricted to the wet "hot spots" where the greatest concentrations of larvae are found.

On the other hand, some investigators have felt that certain insecticides could be used advantageously on poultry droppings. That diazinon, ronnel, Bayer 38156, and malathion, when applied to poultry manure, were effective against fly larvae and relatively nontoxic to predatory mites. Control of the house fly in poultry manure by these mites, under semifield conditions, ranged from 86 to 99 per cent, depending on the species of mite. The use of sugar-based baits against adult flies was another approach to the integrated control programme. Also, cockerels were allowed to run freely under caged hens, and the only food they obtained was what they could find, primarily fly larvae.

Vinegar Flies

These very small flies are cosmopolitan, and are usually known as vinegar flies. They are so small that they can pass through ordinary "flyscreen," and are therefore common pests in homes, restaurants, fruit markets, canneries, cellars, or basements containing fermenting liquids (vinegar, cider, wine, beer), and in other places that contain attractive food materials.

Description. All the economically important species are superficially similar in appearance. They are dull brownish yellow or brownish black in their over-all aspect. A feathery bristle on the antenna helps to distinguish *Drosophila* from other small flies. The adult of the best-known species, *D. melanogaster* Meigen, is 3 mm long, with red eyes, tan-coloured head and thorax, and the abdomen is blackish on top and grayish underneath. Other species that commonly become pests.

Life Cycle. Vinegar flies lay their eggs near the surface of fermenting or rotting fruit and vegetable material or in unclean garbage cans. The pearly-white, cylindrical eggs, which can be observed with a hand lens, have 2 to 4 threadlike filaments at one end, and these filaments protrude above the liquid material in which the eggs are laid. The tiny maggots crawl out of the liquid food to some drier substance, preferably loose soil, to pupate. The brown, seedlike pupae have 2 hornlike stalks at the anterior end. About 500 eggs are deposited by a single female, on an average, and since the life cycle may require only 8 to 10 days in warm weather, the reproductive potential is enormous.

Control of Vinegar Flies

The most important aspect of *Drosophila* control is the elimination of food and breeding sources. This refers not only to the most obvious materials, such as piles of rotting and fermenting fruits and vegetables, but also to such sources as peelings accidentally swept under appliances or counters, discarded ketchup or milk bottles that have retained a portion of their contents, dishwater or water from floor mappings that has seeped into crevices, garbage-laden water from sinks, or a foul, soured mop. Vinegar flies constitute a serious problem in rural communities that are close to extensive sources of the insects, such as tomato fields or canning plants. Cannery sanitation is extremely important. Egg deposition in boxes of tomatoes in the field or on pallets in the receiving stations of canneries can be prevented for a 24-hour period by the application of pyrethrum dust, principally because of its repellent effect.

Piles of culled fruit are important sources of vinegar flies and driedfruit beetles. These pests can be controlled by suitable insecticides, such as liquid sprays or granular formulations. Insecticide granules have to be mixed into the mass of fruits if the pile is more than 24 in. (60 cm) deep. Larvae can be prevented from developing long enough for the fruit wastes to become too dry to attract insects. Sprays of azinphosmethyl, fenthion, or trichlorfon, or granular formulations of malathion, endrin, or heptachlor have proved to be effective. For control of vinegar flies in wine cellars, a thermal aerosol of 0.25' per cent pyrethrins, combined with either piperonyl butoxide or Tropital as synergists, was found to be effective, but treatments once or twice daily were required to keep the insects under control.

Substantial reduction in vinegar fly populations in tomato fields was obtained by releasing large numbers of the adult flies of both sexes that had been sterilized with 1 per cent aqueous apholate, a well-known insect chemosterilant. The investigators stated that release of sexually sterilized flies in a programme such as they described "shows promise of controlling *D. melanogaster* in commercially grown tomatoes, especially if the programme is conducted on a community-wide basis".

In the home, space sprays, aerosols, or dichlorvos resin strips, as they would be used for house fly control, have been effective against vinegar flies.

Chapter 11

Arthropod Pests

More than a million species of animals have been discovered and named. They have been divided into the most inclusive of the taxonomic groups, called *phyla* (singular *phylum*) from the Greek *phylon*, meaning tribe or race. For examples, animals with a backbone, insects, snails, starfish, earthworms, roundworms, flatworms, jellyfishes, sponges, and amoebae are members of different phyla.

The members of each phylum have certain common characteristics. The largest phylum, Arthropoda, includes animals having a segmented body, bilateral symmetry, paired, jointed appendages, usually terminating in claws, a chitinized exoskeleton secreted by the epidermis and molted periodically, striated muscles, a tubular alimentary canal, a ventral nervous system, and a dorsal heart.

Within a phylum there are smaller groups of animals called *classes*, which also have common characteristics that distinguish them from other groups. This is evident by merely recalling the common names of representative animals in the different arthropod classes, such as crayfish, crabs, lobsters, and sowbugs (Crustacea); centipedes (Chilopoda); millipedes (Diplopoda); spiders, mites, ticks, scorpions, and harvestmen (Arachnida); insects (Hexapoda); and several others; each group brings to mind a distinct image.

Some of the arthropod classes contain economically important pests, and will therefore be briefly characterized in the present section merely to present in one place an over-all view of the phylum. They will be discussed further, in detail depending on their importance. Within a class there are groups of animals called *orders*, with common characteristics that distinguish them from other groups. Again, the common names of representative groups of animals give a good idea of what is meant by the term.

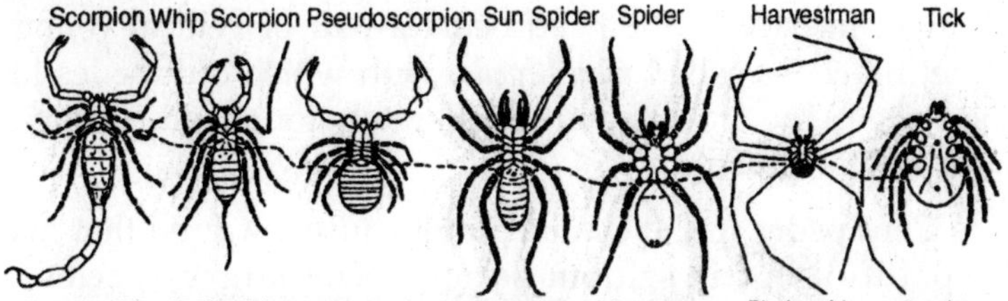

For example, in the class Arachnida are the spiders and related orders: scorpions, whip scorpions, pseudoscorpions, sun spiders, harvestmen, and the ticks and mites. In the class Hexapoda are such important orders as silverfish, cockroaches, crickets and their allies, termites, thrips, lice, bugs, butterflies and moths, beetles, flies, fleas, and the bees, wasps, and ants. Each order has clearly evident similarities among its *families*, and each is different in important ways from the other orders.

Directional Orientation

Certain terms are used in biology to designate the various regions of an organism. The following terms will be found to be useful: *anterior*, at or toward the front; *posterior*, at or toward the hindmost part; *dorsal*, of or pertaining to the upper surface; *ventral*, of or pertaining to the undersurface; *proximal*, the part of an appendage nearest the body; *distal*, the part of an appendage farthest from the body; *caudal*, pertaining to the posterior or anal extremity; also, many words, such as lateral, marginal, etc., which are self-explanatory.

In description, one may employ adjectives formed from such words as thorax, abdomen, and pleuron—thoracic ganglia, abdominal appendages, and pleural regions.

Class Chilopoda (Centipedes)

The centipedes are the closest relatives of the insects for, like the insects, they have a single pair of antennae, breathe by a series of tracheae, and the reproductive organs open at the posterior end of the body. The body is flattened, and consists of 15 to 173 segments, each with a single pair of legs except the last two and the one just behind the head. The segment in back of the head bears a pair of ventral venom-secreting claws called *maxillipeds*, with which centipedes kill their prey. The antennae are moderately long, and consist of at least 12 segments.

Centipedes in the United States cause a wound that may be painful, but is seldom serious. The larger species of *Scolopendra* in tropical countries may reach 30 cm in length, and the amount of venom issuing from their claws is very painful and even dangerous to man.

Class Diplopoda (Millipedes)

Millipedes differ from centipedes in that almost every segment bears 2 pairs of legs, the body is cylindrical instead of flattened, there are no venom-bearing claws, the antennae are short, and the reproductive organs open far forward on the body. Millipedes crawl slowly, whereas centipedes crawl rapidly. On the seventh segment of the male, one or both pairs of legs are modified as copulatory organs. A pair of scent glands on each segment, opening laterally, discharge a repugnant secretion. In some species, it can be discharged to a distance of nearly a meter and, in the case of some tropical species, has caused blindness in children.

Class Arachnida

Most of these arthropods are terrestrial forms, although a few mites are aquatic. They have tracheae or book lungs for breathing, simple eyes (ocelli), no antennae or wings, and have 4 pairs of thoracic legs. They have a chitinous exoskeleton, and the body is divided into 2 portions—an unsegmented cephalothorax and an abdomen. The cephalothorax may be broadly joined to the abdomen or joined by a narrow pedicel.

Arachnids, have 6 pairs of cephalothoracic appendages. The first pair, the *chelicerae,* are either 2- or 3-segmented, and may bear pincers for holding and crushing prey. However, they may also serve as venom-bearing fangs, intromittent organs for transferring sperm, or spinnerets. The second pair of appendages are 6-segmented, and are called *pedipalps.* These may also serve as intromittent organs, as pincers, for crushing food, as sensory organs, or for other functions.

Arachnids have 4 pairs of walking legs, usually 7-segmented. Each leg is composed of a *coxa, trochanter, femur, patella, tibia, metatarsus,* and *tarsus.* Most arachnids are carnivorous, but because they have no mandibles, they commonly predigest their food in the mouth and then suck the juices into the stomach. Some species crush their food with their chelicerae. The black widow spider cannot crush its prey, but pierces it and injects a digestive fluid, then sucks out the digested food. Spiders breathe by means of "book lungs" or tracheae, or sometimes both. Arachnids are oviparous, except for the scorpions and certain mites. There are 11 orders of living arachnids, of which 7 will be considered of more or less importance as household pests, although some are harmless and are of interest only because they may be confused with bona fide pests.

Order Araneida (Spiders)

Whereas the body of an insect is divided into 3 parts—head, thorax, and abdomen—as has just been noted, the body of the spider is divided into 2 parts, the head and thorax being intimately fused into a cephalothorax and the abdomen being usually soft, rounded, and unsegmented. The cephalothorax and abdomen are separated by a narrow pedicel, as is also true of some other arachnids, including the solpugids and whip scorpions. The cephalothorax of spiders is provided with a dorsal, hardened shield or carapace, which bears the simple eyes (ocelli) at the front end. The eyes are usually arranged in 2 rows of 4 each, but there may be fewer eyes or none at all.

In the spiders, the female pedipalps are short and leglike, but in the males they have become modified to form copulatory organs. The last segment is enlarged and knoblike, and is used for transfer of sperm. The spiders are the only arachnids in which the chelicerae are provided with venom glands that open at the tip of the fang. In some spiders, these glands are located entirely within the chelicerae, and in others they extend backward into the first section of the cephalothorax.

The organs of respiration consist of leaf-like external book lungs, or book lungs and tracheae. Most spiders have 1 pair of book lungs, but tarantulas have 2 pairs. They open on the ventral side of the abdomen as slit-like clefts. If a tracheal system is present, the spiracle is usually located just in front of the spinnerets.

Most spiders spin silken webs. The webs are spun from as many as 6 spinnerets, which are transformed vestigial legs and are located ventrally near the tip of the abdomen. Webs may serve as nests or as traps for prey. A spider's silk must serve the dual purpose of transporting her to any part of the web, and also of trapping insect prey.

She may have to secrete 2 kinds of web, dry and sticky, supplied by different glands in her abdomen. For example, in the orbweaver's web, the central "lookout," where she spends most of her time, and also the "spokes" that radiate to the outer rim of the web, consist of dry webbing over which she can run rapidly, upside-down, aided by hooks, toothed claws, and bristles at the ends of her legs. She avoids the sticky threads that join the spokes and are for the purpose of catching the insect prey. She can perceive the slightest disturbance of any part of the web, her highly developed tactile sense compensating for her poor eyesight. After reaching her prey, the spider quickly envelops it in fresh threads, and bites it with her sharp upper jaws, injecting a venom that quickly kills it.

After the prey dies, the threads of the web entangling it are severed, and the prey is transported to the central "lookout" and hung up by a short thread. To consume her victim, the spider bites it, meanwhile secreting a digestive fluid, then sucks up the dissolved contents together with the digestive fluid. Dissolved muscles and intestines of the prey are imbibed within a few hours, and the indigestible remnants are dropped out of the web.

Webbing may also be used for transportation when it is borne aloft by air currents. The interesting observation that although spiders were wingless, a young spider (spiderling) could float its threads on the breeze, often reaching tremendous heights and sailing for long distances, thus colonizing new areas at a rate not possible even for winged insects. He stated that the rigging of ships 320 km from the nearest land had been "showered with tiny aeronauts riding on silken streamers."

Some spiders do not spin true webs, but live in burrows or crevices (wolf spiders).

The courtship and mating of spiders is so strange and complicated as to be almost beyond belief. They are described in great detail in W. J. Gertsch's fascinating book, *American Spiders*. Soon after he becomes mature, the male spider transmits some seminal fluid to a delicate web which he has spun. He then collects the sperm and stores it in bulbs of his pedipalps (palpi) until the opportunity for mating arises. The sperm-induction procedure in the mating of black widow spiders and the courtship and copulatory behaviour of that species, lasting over 2 hours. Most bizarre is the sperm-induction sequence practiced by tarantulas. It is so complicated and replete with ritual that it sometimes requires 3 or 4 hours.

Spiders are predators, mostly on insects ensnared in their webs. A few species are very venomous. In the United States, the bite of the black widow spider and the brown recluse spider or related species may be very painful, and can lead to serious general systemic disturbances, or even death. The aid of a physician should be obtained as soon as possible.

Order Acarina (Mites and Ticks)

The body divisions of the Acarina do not correspond to those of other arachnids, for the head, thorax, and abdomen are fused together, forming an unsegmented body. This fact has resulted in a special terminology.

The *gnathosoma* or *capitulum* represents the first 2 primitive segments, and bears the chelicerae and pedipalps. The *idiosoma* includes the rest of the body. Many mites have transverse, fine lines, but these have no relationship to segmentation. The appendages of most Acarina are somewhat similar to those of spiders, except that the newly hatched young have only 3 pairs of legs. (Mites of the family Eriophyidae have only 2 pairs.) Most mites have a tracheal system opening through 1 to 4 pairs of spiracles, located near the bases of the appendages in the anterior half of the body. Some species have no tracheae, and obtain oxygen by diffusion through the cuticle.

The majority of species are oviparous (egg-laying), but some are ovoviviparous, the eggs hatching within the body of the female. There are 4 stages in the development of mites: the egg, the hexapod larva, the octapod nymph, and the adult. In each postembryonic juvenile stage, there may be 1 or more instars. Most mites are very small, and many species are pests of plants or stored food products. Other species are parasites of man and animals, and some are disease vectors or are responsible for allergies and asthma.

Order Scorpionida (Scorpions)

Like the spiders, scorpions have 2 body regions, the *cephalothorax* or *prosoma*, and the *abdomen*. However, the abdomen is segmented, and consists of a large, anterior preabdomen or *mesosoma* of 7 segments and a long, narrow, tail-like posterior postabdomen or *metasoma* of 6 segments, terminating in a vesicle or *telson* bearing a venomous stinger. The anterior part of the cephalothorax bears 3 groups of eyes: a median group of 2 eyes and a lateral group of 2 or 3 eyes at each margin. The cephalothorax also bears small, chelate chelicerae, large, conspicuous, pincerlike pedipalps, and 4 pairs of legs. The general body form and thornlike rigid stinger of the scorpion identify it for most people. It ranges throughout the southwestern United States and northern Mexico, and is 11 to 12 cm long—the largest scorpion in the United States.

Scorpions breathe by means of 4 pairs of book lungs in the ventral blood sinus. Paired coxal glands, opening in the coxae of the third pair of legs, provide for excretion. They do not lay eggs, but give birth to active young. These climb upon the back of the mother and remain fastened by their pincers until after the first molt, nourished by yolk material stored in their bodies. All scorpions are venomous, but most species are not dangerous. In Arizona, however, there is a species that has caused more deaths over a 36-year period than all other venomous animals of that state, including rattlesnakes.

Order Pedipalpida

In this order, the cephalothorax bears 8, the chelicerae are small, 2-jointed, and chelate; and the pedipalps are strong, and have 6 joints. In many species, the last 2 segments of the pedipalps are modified to form pincers used in seizing prey, tearing it apart, then passing it to the chelicerae to complete the maceration. The first pair of legs are not for locomotion. These legs are very long, and the long, many-segmented tarsi are used as feelers. These harmless arachnids are nocturnal, hiding during the day beneath leaves, rocks, or logs, usually in damp places, although there are a few desert species

There are 2 suborders. In the Uropygi, there are 2 glands that open one on each side of the anus and discharge a stream of fluid when the animal is irritated. This fluid has the odor of vinegar, which suggested to the Spanish-speaking settlers of the Southwest the name *vinagrn,* later modified to "vinegarroon" in English. The last abdominal segment bears a *flagellum,* which is very long with some species. The body length may range from only 2 mm to as much as 6.5 cm, as in *Mastigoproctus giganteus* (Lucas), which occurs in the southern United States from coast to coast. *Mastigoproctus giganteus* gets water from its prey and by drinking from a moist substrate which is located by the sensitive, antenniform forelegs. The whip scorpion probably does not have a protective epicuticular wax layer, such as that possessed by insects and many other arthropods, to protect it from desiccation. At 26°C, it readily loses water to the air at relative humidities up to 95 per cent.

Order Chelonethida (Pseudoscorpions)

The pseudoscorpions are relatively small, ranging from 4 to 10 mm in length. They resemble scorpions, but have no postabdomen or stinger, and are harmless. The abdomen is 11-or 12-segmented, and is rounded posteriorly. Their conspicuously large pedipalps bear lateral claws called chelae, with a poison gland for disabling their prey. Unlike scorpions, the pseudoscorpions do not possess book lungs. They have 2 pairs of ventral tracheal spiracles on the third and fourth abdominal segments. Like scorpions, they have a pair of glands that open on the coxae of the third pair of walking legs and provide for excretion.

Order Phalangida (Harvestmen)

The harvestmen or daddylonglegs are distinguished most readily by their long, slender legs. When they crawl, the small body appears to be suspended from the long, arched legs. The cephalothorax is broadly joined with the short, segmented abdomen, has 1 pair of simple eyes, and a pair of odoriferous glands. The pedipalpi are leglike. A pair of tracheal tubes opens on the second abdominal sternite. A pair of coxal glands, opening on each side between the third and fourth coxae, provide for excretion.

The harvestmen easily lose their legs or parts of their legs, and cannot regenerate them. They drink frequently, and must have water readily available. They are harmless, and should not be confused with true spiders.

Order Solpugida (Solpugids)

The solpugids, also called "sun spiders," are harmless to humans. They are predaceous arachnids that occur mostly in dry tropical and in warm temperate regions. Of the 200 known species, about 60 are found in the southwestern United States.

Solpugids are most strikingly characterized by their enormous, 2-jointed chelicerae, with the pincers articulating vertically and a rounded abdomen without a tail-like post-abdomen, telson, or stinger. They do not have venom glands. The pedipalps are simple, non-chelate, leglike, and terminate in an adhesive organ. Solpugids seize their prey with their pedipalps and pass it to the chelicerae, where it is crushed.

The macerated and predigested tissues of the prey are then sucked into the mouth and esophagus. Only the last 3 pairs of legs are ambulatory; the first pair is tactile. The abdomen is 10- or 11-segmented. Paired spiracles are located between the second and third coxae and in the third and fourth abdominal segments, leading to a well-developed tracheal system. A pair of coxal glands and a pair of Malpighian tubules provide for excretion.

Class Hexapoda (Insects)

The hexapods comprise the largest and most highly developed group in the phylum Arthropoda. Some authorities estimate that over a million species have already been described, more than all other animal species combined. Insects also occur in incredible numbers. A total of 1018 live individual insects on the earth at any given moment. Adult hexapods may be distinguished from other arthropods by a number of structural characteristics, including a segmented body divided into 3 regions, the head, thorax, and abdomen. Adult hexapods possess 1 pair of antennae, 3 pairs of legs, and most have 2 pairs of wings, although some have 1 pair and others none. No other invertebrates have wings. Furthermore, insects have a front lip (*labrum*), 1 pair of mandibles, 1 pair of maxillae, and a hind lip (*labium*). Their gonopore is posterior, and they usually undergo a change of form (*metamorphosis*).

Respiration in hexapods usually takes place by means of a system of tubes called *tracheae*, that divide and subdivide internally until they terminate in the delicate network of *tracheoles* that reaches every organ, tissue, and cell of the body. Hexapods have a simple, open circulatory system.

The modern tendency is to divide the class Hexapoda into 3 subclasses. Subclass Myrientomata contains 1 order (Protura), subclass Oligoentomata contains 1 order (Collembola), and the subclass Insecta (Euentomata) contains the remaining 23 orders. This modern classification is based on the evidence that the 3 subclasses represent distinct evolutionary lines.

However, since the present work has primarily economic rather than taxonomic objectives, for the sake of convenience all orders of hexapods will henceforth be referred to as "insects" Like-wise, for the sake of brevity and convenience, the older concept of all Hexapoda being divided into 2 subclasses, Apterygota and Pterygota will be retained.

The Exoskeleton

The skeleton of an insect is external. It forms a shell-like support, and provides protection for the internal organs. This outer shell is light, but very strong. This is demonstrated by the tremendous impact that an insect can withstand without injury—as, for example, when colliding with a solid, stationary object during flight. The hard material on the exterior comprises a thin covering of the entire insect known as the *cuticle.* It is secreted by a single, continuous layer of living cells, the *epidermis* (sometimes called "hypodermis"), which is supported on a noncellular *basement membrane.*

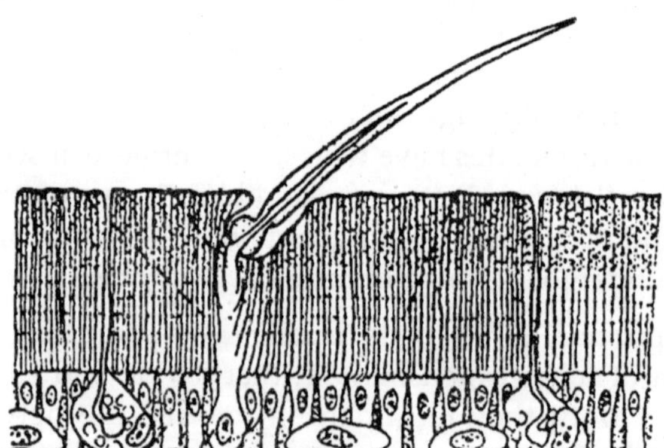

Fig. Exoskeleton

It was once assumed that the polysaccharide *chitin* was the main constituent of the cuticle of insects and other arthropods, and such terms as "chitinous exoskeleton" are often found in fairly recent literature. However, although chitin may comprise as much as 60 per cent of the cuticle in a few cases, it may also be present in amounts as small as 1 or 2 per cent. Arthropod cuticle may be thought of as a continuous protein layer that usually contains chitin within and waxes on the outer surface. A review of the literature on the structure of insect cuticle.

The *procuticle* comprises the bulk of the cuticle, and is immediately above the epidermis. It is essentially a hydrophilic chitin-protein complex containing a considerable quantity of water. The outer layer of the cuticle is called the *epicuticle* This consists of a proteinaceous *cuticulin* about 1in thickness, containing no chitin, and covered with a layer of lipid, usually a hard wax, averaging about 0.25 1u in thickness. This wax, and particularly 1 or more packed and oriented monolayers of it at the cuticulin-wax interface, is the principal barrier to water loss. After some insects molt, dermal glands secrete a fluid that spreads over the surface of the wax layer to forma "cement" that appears to be similar to shellac. Until somewhat recently, it was not known how the wax was brought to the surface of the cuticulin.

Arising from the epidermal cells and extending outward through the procuticle of most insects, and sometimes into (but not completely penetrating) the overlying cuticulin are minute ducts called *pore canals.* By means of the electron microscope, it was shown that these canals terminated in wax *canals* that branched out and reached the outer surface of the cuticulin, depositing a liquid wax similar in appearance to the liquid crystalline phases of lipid water systems. By means of sorptive dusts, a sufficient amount of wax can be removed from the solidified epicuticular wax layer to cause a lethal rate of water loss, providing means of controlling some insects by nonchemical methods.

Setae (bristles or hairs) are sensilla that arise from pits in the cuticle surface in a sort of ball-and-socket arrangement. Setae are formed by 2 cells: a sensillum-forming cell (*trichogen*) and a socket-forming cell (*tormogen*), a, i, k) and, of course, whatever cells that give rise to the sensory nerve(s). In the pit, the cuticle (articular membrane) is relatively thin, and so is the cuticle of the seta, etc., particularly in the case of some of the chemoreceptors. Cuticles of olfactory sense hairs are perforated by many minute pores. Cytoplasmic filaments, arising in the trichogen and tormogen cells, pass up through the cuticle to enter the seta, forming a cytoplasm continuum in close contact with chemicals that may be deposited on the surface of the cuticle, and thereby forming one of the pathways for entry of insecticides through the integument.

The insect cuticle does not readily decompose, but retains its original form after the insect dies, which makes possible the preservation of insects for hundreds of years with their original lifelike appearance. Thus, the mounting and preservation of insects are comparatively easy.

An insect's exoskeleton cannot be readily dissolved away by strong acids, alkalis, or solvents, but fortunately for man's attempts to control insects by means of contact poisons, it can be penetrated by solvents, surface-active materials, vapours, and gases. In fact, it can be penetrated by organic compounds of all kinds, even when they are solids and are not dissolved in a liquid solvent. Fortunately, the rate of penetration of organic insecticides into insect cuticle is much greater than into mammalian skin. Even inorganic insecticides can penetrate the cuticles of some insects, such as cockroaches, at a lethal rate. The *spiracles* which are openings for the tracheae, form a channel of entry for liquids of low surface tension, and for vapours and gases.

Insects are divided by constrictions into ring-like plates called *segments* Thin, flexible portions of the cuticula called *conjunctivae* connect the segments. These conjunctivae may be folded into the body of the insect, allowing for expansion between segments much in the manner of an expanding accordion. Other constrictions or infoldings of the cuticle called *sutures* run in various directions on the surface of the exoskeleton. The internal infoldings often provide points for muscle attachments. The areas bounded by sutures are called *sclerites*. The shape and position of the sclerites are often used in the description of species in some insect groups.

In addition, the hairs, spines, setae, ridges, and other exoskeleton structures are used in taxonomy. External structures may be merely projections of the cuticle, or they may be associated with the underlying body structures, such as the sense organs.

The dorsal or upper face of each segment is called the *tergum* the ventral or lower face is called the *sternum* and each lateral face is called a *pleuron*. Each of these faces is made up of 1 or more *sclerites*. The dorsal part of a segment may be called a *tergite*and the ventral part a sternite - terms usually restricted to the abdomen.

If reference is made to the thorax, the term *notum* is generally employed instead of tergum. Thus, one refers to the dorsal part of the first segment of the thorax as the *pronotum* but the corresponding part on the ventral area as the *prosternum*.

Molting

An insect can increase in size during the molting process until the hardening cuticle restricts further expansion. At molting, the old cuticle is largely dissolved by special molting fluid secreted by glands or by the epidermal cells in general. New, soft cuticle is deposited beneath the remnant of the old cuticle. The old, chemically weakened cuticle is split by internal blood and air pressure and muscular action, the split occurring along some specific line of least resistance which differs in its location among the various insect species, but frequently is along the top of the thorax and head. A cockroach (*Blatta orientalis*) is shedding its old skin (black). The insect is white immediately after molting, but becomes black within a few days. Molting appears to be a difficult and labourious process, and is a critical period in the life of an insect. The old cuticle cast off by an insect is called the "cast skin" or exuvia.

The Head

The head appears to comprise a single segment, but embryological observations indicate that it is a fusion of 5 or 6 segments possessed by primitive insect progenitors. The arrangement of the skeletal features of a cockroach's head is fairly well defined and will, therefore, be used as a generalized model. The principal external regions of the head are designated as *vertex, front* or *frons, gena, clypeus,* and *labrum*. The eyes, antennae, and mouthparts are borne on the head.

Eyes

An adult insect usually possesses both compound and simple eyes. The compound eyes are among the most characteristic features of adult insects. They are composed of many small, hexagonal lenses or facets fitted closely together. Each facet is the exposed face of an independent lens and an independent eye unit called an *omrnatidium*. Depending upon the species, there may be from 50 to 30,000 ommatidia in a single compound eye.

Besides the 2 compound eyes, an adult insect usually has 3 simple eyes or ocelli, but may have 2,1, or none, depending upon the species. Simple eyes have only a single facet. The larvae of some orders of insects, such as the flies, beetles, butterflies, and ants, never have compound eyes. They may have from 1 to 6 or more simple eyes on each side of the head. Sometimes, şeveral simple eyes form a cluster. Nymphs of bugs, grasshoppers, and dragonflies may have compound eyes.

Insects cannot change the focus of their eyes (accommodation), and they have no protection equivalent to the eyelids and eyelashes of vertebrates. Protection is afforded by the cuticle that covers the entire insect body, and which continues in an unbroken sheet over the eyes. Over the eye surface, however, the cuticle is thin and transparent, admits and refracts light, and is curved to form a lens.

Antennae

The antennae of insects are extremely variable in form. Various names, for the most part self-explanatory, have been more or less generally accepted to designate the different types of antennae: setaceous (bristlelike), filiform (thread-like), moniliform (beadlike), serrate (sawlike), pectinate (comblike), clavate (clubshaped), capitate (knobbed), lamellate (bearing many plates), and plumose (featherlike). Many forms of the above types, as well as miscellaneous unclassified forms, may be found among the insects.

Antennae may be used as tactile organs, or may possess organs for smelling or hearing. In general, they are used as sensory aids for locating food and finding mates, and it appears that the ants use their antennae to communicate with others of their species.

Mouthparts

Insects are divided into groups of those having chewing (mandibulate) mouthparts and those having various modifications of the primitive chewing mouthparts, including many that are used only for sucking or siphoning.

Chewing Mouthparts. Examples of insects with chewing mouthparts are silverfish, cockroaches, grasshoppers, crickets, termites, earwigs, psocids, biting lice (Mallophaga), dragonflies, scorpionflies, beetles, beetles, ants, sawflies, woodwasps and larvae of moths, butterflies, and primitive flies.

The mouthparts of the cockroach are of a fairly generalized type, and may be used as an example of the chewing mouthparts. The upper lip (*labrum*) and lower lip (*labium*) oppose each other in a vertical plane. The labrum is a sclerite of the head, but the labium is a true mouthpart appendage. The actual biting and chewing appendages are the *mandibles* and *maxillae*. They operate in a horizontal plane, i.e., from side to side. The mandibles are the first pair of "jaws" that can be seen when the labrum and labium are removed. They are strong, heavily sclerotized, often darkened, and possess teeth and grinding surfaces. They are operated by 2 sets of muscles, one set that closes them against each other and another set that pulls them apart.

The second pair of "jaws" are the maxillae. A maxilla is made up of a number of parts. The part modified for tearing food apart is called the *lacinia*. The maxillae bear segmented sensory palps provided with tactile hairs and also organs of smell and taste. In cockroaches, chemoreceptors involved in feeding (taste) are confined to the mouth. The Hemiptera, Lepidoptera, and Diptera also have chemoreceptors on the tarsi, while the Hymenoptera have similar receptors on the antennae as well as on the mouthparts.

The lower lip or *labium* performs a function similar to the lower lip of vertebrates. In addition, it bears a pair of sensory palps whose function is similar to those of the maxillary palps. As might be surmised from the paired palps, the labium was formed by the fusion of primitively separate appendages, the "second maxillae."

The *epipharynx* is a sensory organ attached to the inner surface of the labrum that is believed to contain end-organs of taste and is somewhat analogous to the palate of higher animals. It is of interest principally because it becomes modified into highly specialized structures in some of the sucking insects.

The *hypopharynx* is a tonguelike organ that forms the floor of the mouth cavity. The salivary duct opens in a pocket between the hypopharynx and the labium. The mouth opens between the hypopharynx and labrum.

Various Types of Sucking Mouthparts. Chewing mouthparts are the most primitive type, from which the various sucking types have evolved to form various interesting structural modifications. Sucking mouthparts are thus thought to be homologous with chewing mouthparts, even though they usually differ greatly both in appearance and function. The following types of sucking mouthparts among adult insects:

1. The *rasping-sucking* type of mouthparts is possessed by the thrips. This type is considered by some authorities to be intermediate in structure between the chewing and piercing-sucking types. The mouthparts of thrips are somewhat asymmetrical, the right mandible being vestigial. The styletlike left mandible and the maxillae can be protruded through a circular opening at the conical apex of the head. To extract plant juices, a thrips thrusts its head downward and backward in the manner of a pickax, the stylets piercing the leaf or fruit cuticle. The plant sap is then withdrawn through the mouth cone which, when closely pressed to the plant surface, can probably provide considerable suction.
2. The *piercing-sucking* type of mouthparts is characterized by a tubular, usually jointed, beak, which encloses the stylets The stylets are modifications of the mandibles and maxillae of the chewing insects, sometimes supplemented or replaced by the labrum-epipharynx and hypopharynx. There are several subtypes of the piercing-sucking mouthparts, namely, the "bug or hemipterous subtype"; the "louse or anoplurous subtype"; the "common biting-fly or dipterous subtype [including in the latter the mosquitoes and horse flies; the "special biting fly or muscid subtype" [including the horn fly and stable fly and tsetse fly; and the "flea or siphonapterous subtype". The successive stages in the insertion of the feeding bristles (mandibular and maxillary bristles) of the mouthparts of a bug.
3. The *sponging* type of mouthparts is illustrated by the house fly, blow flies, and others. In such flies, the end of the fleshy, elbowed, and retractile labium is specialized into a pair of large, spongelike organs, the *labella*. The labella contain many furrows or

channels called *pseudotracheae,* through which liquid material is drawn, possibly by capillarity. Liquid food is drawn up from the pseudotracheae through the food channel into the esophagus. It is usually n-possible for insects with mouthparts of this type (nonbloodsucking Muscidae, Syrphidae, and many other Diptera) to pierce the skin, but some have labellar teeth with which they can abrade the skin, e.g., the eye gnats (*Hippelates*).

4. In the *siphoning* type, found in the adult Lepidoptera (moths and butterflies), the proboscis is derived almost entirely from the galeae of the maxillae of the more primitive mandibulate insect. The 2 elongated maxillae fit together to form a tube which, in the resting position, is coiled and held close to the underside of the head. The labial palpi are often large and conspicuous. Except for a few species, notably the fruit-piercing moths of some tropical countries, moths and butterflies cannot pierce the epidermis of animals or plants. They usually feed by uncoiling the proboscis and sucking exposed liquids, such as nectar, through it.
5. In the *chewing-lapping* type of mouthparts, illustrated by the bees and wasps, the mandibulate origin of the labrum and mandibles is readily recognizable. The mandibles may be modified to serve as tools for a variety of tasks performed by the various species possessing such mouthparts. The maxillae and labium are elongated, forming a long tongue which is of great importance in feeding.

The type of mouthparts possessed by the adult insect does not necessarily indicate the type possessed by the young of the same species. Because of their gradual or incomplete metamorphosis, in the majority of the more primitive orders of insects the adults have mouthparts of much the same appearance and function as those of the immature forms. The principal orders of this type are the Protura, Thysanura, Collembola, Orthoptera, Isoptera, Dermaptera, Psocoptera, Thysanoptera, Phthiraptera, and Hemiptera. (However, in the Coleoptera, with complete metamorphosis, larvae and adults have chewing mouthparts.) In the Odonata, Trichoptera, Hymenoptera, and Diptera, the mouthparts of the immature insects usually differ markedly from those of the adults. In the Lepidoptera, the mouthparts of the larvae are always different from those of the adults, the larvae having mandibulate and the adults having haustellate (sucking) mouthparts.

The cocoon or attachments for the pupa of Lepidoptera are formed of silk threads spun by silk-spinning organs in the mouth of the larva. Adult Neuroptera and Coleoptera have biting mouthparts, and the immature life stages have mouthparts similar to those of the adults except that in some species (antlions, lacewing larvae, and diving beetles) the mandibles and maxillae are grooved to facilitate the sucking up of the liquids of the insects' prey after the body has been pierced.

The Thorax

The second of the 3 body regions, called the *thorax*, is composed of 3 segments the *prothorax*, the *mesothorax*, and the *metathorax*. In the grasshopper and the cockroach, only the pronotum is visible from above; the wings cover the mesonotum and metanotum. Each thoracic segment bears a pair of legs, and in the case of winged insects, the mesothorax and metathorax bear the wings. Also, in the case of adult insects, the latter 2 segments usually each bear a pair of spiracles. The locomotory appendages of adult insects are thus confined to the thorax, but among the immature insects, locomotion may be aided by 2 to 8 pairs of fleshy, unjointed *prolegs* on the abdominal segments, or locomotory appendages may be entirely absent.

Legs

Three pairs of jointed legs are practically always present in adult insects, and are generally present in the immature stages. Nearly all adult animals having 6 legs are insects [Some adult mites of the family Tenuipalpidae (*Larvacarus* and *Phytoptipalpus*) and the males of some Podapolipodidae also have only 3 pairs of legs. (The females are frequently legless.)], and the name of the class, Hexapoda, is based on this fact. The 5 parts of the legs are held together by intersegmental membranes at the joints, and are operated by internal muscles. The 5 parts, beginning next to the thorax, are called *coxa, trochanter, femur, tibia,* and *tarsus*. These terms should be remembered, because they are often encountered in entomological literature. The trochanter in some hymenopterous species has 2 segments, and the tarsus may have 1 to 5 segments.

There are usually 2 claws on the tarsus, and there are frequently pads (*pulvilli*) beneath the claws that offer greater purchase against smooth objects. The ability of the pulvilli to cling to smooth surfaces is usually enhanced by many small hairs which sometimes, as in the case of the house fly, exude a sticky substance. Between the claws, there may be another type of pad or lobe called the *arolium* which has considerable suction force, and enables the insect to cling to objects with little or no aid from the claws. When the arolium is "hairlike," as in flies, it is called an *empodium*. Among the thrips (Thysanoptera), the claws are very small or may be entirely absent, and the tarsus terminates in a large bladder like arolium. Among the scale insects, sucking lice, and some biting lice, there is only a single claw on the tarsus. The claws and the associated structures at the tip of the tarsus are presently considered to be a separate segment, the *pretarsus*.

Depending on the mode of life to which an insect has become adapted, great variation can be seen in the morphology of the legs of different insect species. A rapidly running insect like the tiger beetle (Cicindelidae) will have long, slender legs, while an insect that leaps, such as a grasshopper, flea, or flea beetle (*Haltica*) will have enlarged femora or coxae, at least on the hindlegs. The legs of insects that dig into soil or wood are usually short and stout, and may be armed with spurs. The forelegs may be specialized for the grasping of prey, as in the case of the praying mantis and ambush bugs. Aquatic insects often have legs adapted for swimming. Arboreal insects, such as the cerambycid beetles, may have greatly expanded and hairy tarsi and strongly developed claws and pulvilli to afford the maximum grip.

Wings

Among invertebrates, only insects have wings, and insects are the only animals with 2 pairs of wings. Wings, like the 3 pairs of legs, are features which can be used to separate most insects readily from the many small invertebrate species with which they could be confused. Only adult insects have functional wings [The mayflies, before they become adults, molt their wings once.]. Some groups (proturans, springtails, and silverfish) appear never to have had wings in the course of their evolutionary history, while others (lice, bed bugs, fleas, some aphids, and ants; females of the mealybugs and scales; and certain lepidopterous and coleopterous families) have no wings, but had winged ancestral forms. The absence of wings in this latter group of insects does not indicate a primitive evolutionary status, but on the contrary, suggests a high degree of morphological specialization for their particular modes of existence.

Most insects can flex their wings against the top of the body while they are resting, and also fold their wings if they are broad. It is usually the broader hindwings that are folded, and if they are too long, the tips may be folded under so as to be protected by the forewings. Dragonflies and damselflies (order Odonata) and mayflies (order Ephemerida) lack the mechanism for flexing their wings, which are therefore held outstretched laterally or held vertically over the body. This is considered to be a primitive condition, but some of the higher insects, such as butterflies, have secondarily lost the ability to flex their wings.

Typically, there are 2 pairs of wings: one pair on the mesothorax and another pair on the metathorax. The posterior pair, in the case of the true flies (Diptera), is replaced by vestigial appendages called *halteres*, which are believed by some entomologists to function as balancing organs. In the male scale insects, the hind pair of wings is replaced by 2 spines, 1 on each side, which hook into the weakly developed pair of wings of the mesothorax, giving the forewings added rigidity. Among the Coleoptera, the mesothoracic wings (*elytra* singular *elytron*) are horny and shieldlike. When in flight, most beetles hold their elytra to each side like the wings of a glider, depending on the membranous second pair of wings for propulsion. The second pair of wings may be longer than the first pair, but are folded away under the elytra when the beetles are not flying.

The exoskeleton and musculature of the thorax are highly developed to suit the needs of the great muscular activity involved in flight. This is most conspicuous among Odonata, Diptera, and aculeate Hymenoptera, and these insects are among the most active of the insect fliers. The wings of flies and bees move so rapidly that they become invisible.

The wingbeat frequency of one species of midge (*Forcypomyia*) has been recorded at 133,-080 beats per minute, or 0.00045 second per wingbeat.

An examination of the wing pads of a nymph of the lower winged insects or of the chrysalis of Lepidoptera will reveal that an insect's wing is simply a double-layered hollow sac folded out from the body wall, in the lumen of which are nerves, tracheae, and body fluid. The double-layered nature of the wings may also be easily demonstrated by placing small, adult beetles in a damp location until the wings have absorbed enough water to become bloated, thus revealing their saclike construction. The colour of the wings of Lepidoptera depends on the colour and arrangement of myriads of shinglelike "scales" attached to their wings. The shingled arrangement of the scales strengthens an otherwise membranous structure that would be inadequate for rapid flight. Each scale is often strengthened by minute corrugations that diffract light rays and result in the brilliant colour and iridescence usually noted in butterflies. Each scale has a pedicel at its base which serves as an anchor point. The pedicel is enlarged in the middle, and forms the "ball" that fits, in a fixed position, into the "socket" cavity in the cuticle of the wing membrane.

The Abdomen

Most of the segments of the third region of the body, called the *abdomen*, are somewhat similar in appearance. The tergites and sternites are joined by thin, membranous conjunctivae in the same way that successive abdominal segments are joined. The conjunctivae permit expansion, and the abdomen of a gravid female may be distended to fantastic proportions, as in the case of the termites.

Adult insects usually do not have any abdominal appendages, except for those called *genitalia*, used in reproduction. The genitalia are under the heading, "Reproductive System." *Cerci* and *caudal filaments* are borne on the eleventh segment of some insect species. If 11 abdominal segments are present, the respiratory spiracles normally occur on the first to eighth segments. In many species of insects, a fusion of posterior abdominal segments takes place, and it may be impossible to recognize more than 5 or 6 segments.

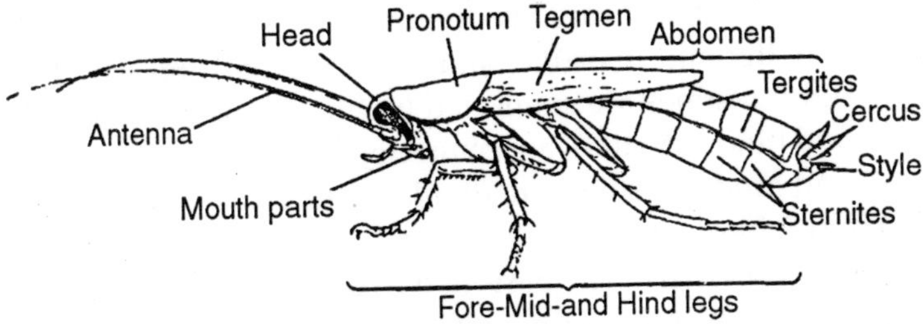

The larvae of some insects have fleshy, unjointed projections of the abdomen, known as *prolegs*, which are used as legs and are considerable aids to locomotion for some of the more elongate forms. The larvae of Lepidoptera never have more than 5 pairs of prolegs, but those of certain Hymenoptera, the sawflies (Tenthredinidae), may have 6 to 8 pairs.

Orthopterans, such as grasshoppers, locusts, katydids, and crickets, that produce stridulatory sounds, have well-developed, thin, membranous tympanic organs for detecting these sounds. Receptive tympanic sensilla are located on the front tibiae or the first tergite of the abdomen. Insects that lack such complex tympanic organs, such as cockroaches, can detect air movements by means of sensilla that may cover the cerci. The cerci of the American cockroach have a partly auditory function.

Internal Anatomy

The cockroach will again serve as a generalized form of insect, this time to illustrate the internal anatomy of the subclass Insecta.

Digestive System

As already stated, the exoskeleton may be likened to a shell serving as an outer skeleton. The alimentary tract can be thought of as a tube within this outer shell, extending toward the rear from the mouth to the anal opening. This analogy is more striking in the case of the larvae of the higher insects. Between the exoskeleton and the alimentary tract is the body cavity, or *hemocoele*.

The alimentary canal is divided into 3 regions: the fore-intestine, the mid-intestine, and the hind-intestine. The fore- and mid-intestines are connected by the *cardiac valve*, and the mid- and hind-intestines are connected by the *pyloric value*. The fore- and hind-intestines are divided into several more or less distinguishable parts, which are well illustrated by the dissected cockroach. The parts of the cockroach's fore-intestine are the *pharynx*, the *esophagus*, the *crop* and the *proventriculus*, or *gizzard*. The gizzard occurs in insects that eat hard substances, and it appears to have a grinding and straining function like the gizzard of birds. The crop serves as a food reservoir, and is absent in some insects, or may be represented as a side sac, or *diverticulum*. The pharynx is a well-muscled pump in insects having sucking mouthparts.

The mid-intestine or *ventriculus* is not differentiated into regions, but a variable number of evaginations of variable size and shape and in various locations arise from it. These are called *enteric* or *gastric caeca;* and digestive enzymes may be produced in them, or they may represent additional digestive surface.

In the hind-intestine of the cockroach, the *ileum*, the *colon*, and the *rectum* may be distinguished. *Malpighian tubules* discharge into the anterior end of the hind-intestine at the pyloric region. The tubules vary in number and length, but the point at which they are attached to the alimentary canal can always be used as a means of establishing the point of juncture between the mid-intestine and hind-intestine. The functions of the Malpighian tubules are primarily to remove the waste products of metabolism (water and uric acid) from the blood stream, empty them into the hind-intestine, and adjust the ionic concentration of the blood (homeostasis).

The length and complexity of the alimentary tract of an insect species are usually related to the kind of food the insect eats. Long and complex systems are found among the herbivorous insects, while those of the entomophagous species and the insects feeding on concentrated food are relatively short and simple.

Circulatory System

The "heart" of an insect, stated in simplest terms, is a flexible dorsal tube, closed at the posterior end and open at the anterior end, with small, valvelike openings or *incurrent ostia* (sing., *ostium*) located along the sides of successive chambers. Blood enters the heart through the ostia as the alary muscles dilate the heart chamber, but the ostia close when circular muscles of the chamber contract, and their valvelike arrangement prevents the blood from returning to the *hemocoeles* or posterior chambers. Because the contractions start at the posterior chamber of the heart and pass forward, the blood is pushed anteriorly, passing through the *aorta*, a constricted portion of the anterior part of the heart. The blood is pumped from the aorta into a series of cavities or sinuses and veins throughout the insect.

The blood bathes the "brain," and from there it passes posteriorly through the hemocoeles, bathing the internal organs and passing into the appendages and wings. The circulation of the blood is aided by the contraction of a *ventral diaphragm*, and often by *accessory hearts* at the bases of legs and wings. One function of the blood is to carry digested food materials from the alimentary canal to all tissues of the body. In contact with the alimentary canal, the blood picks up the products of digestion by osmosis. While the blood is bathing the network of Malpighian tubules, waste products are removed. Insect blood has no red blood corpuscles and plays no part in oxygen transport, although the plasma of chironomid larvae does contain hemoglobin which may aid in respiration.

In the greater number of cases, both oxygen and carbon dioxide are exchanged directly with the external environment through the tracheal system.

Nervous System

In common with all arthropods, the nervous system in the body of insects is located ventrally. Primitively paired *ganglia* in each segment are usually fused together, as are also the double cords connecting the ganglia. In the head, above the esophagus, lies the supraesophageal ganglion or *brain*, linked to the smaller subesophageal ganglion by paired connectives which pass around the esophagus.

Since so much of the nervous activity of insects is on a simple receptor-effector basis, the brain does not serve as an important coordinating centre as it does among the vertebrates. If the brain of an insect is removed, it may continue to live for a considerable period, and can carry on many reflex activities. If a bee is captured and confined in a transparent container when returning to the hive with a load of nectar or pollen, it will die from stress in 2 or 3 hours. However, the excised abdomen will live for 2 or 3 days, and has been found to be an excellent subject to monitor physiologic response to injected drugs or insecticides, free of psychogenic infuences, by means of a recently developed biosonic analyser.

Respiratory System

The *spiracles* are the external openings of the air tubes, known as *tracheae,* which divide and subdivide as they penetrate to all portions of the body. The tracheae eventually divide into microscopic tubes, usually less than 1 in diameter, called *tracheoles*. These reach every cell in the body. The cells obtain oxygen and give off the waste gases of metabolism by diffusion through the walls of the tracheoles.

By microscopic examination of insect tissue, the presence of the tracheae can always be determined by the fact that they are lined with cuticula in fine, spirally arranged thickenings called *taenidia,* which prevent the collapse of the delicate tracheal tubes.

Many insects, for example, parasitic insects spending their larval lives in the bodies of their hosts, respire through their cuticles, even though they may have a tracheal system. Some insects may live for long periods without oxygen. An oestrid larva remaining alive 17 days immersed in oil. Specimens of California red scale, *Aonidiella aurantii,* have lived as long as 40 days after being placed on glass slides, covered with refined spray oil, and then sealed with glass cover slips. The majority survived under these conditions for 10 days or more. If the cover slips were not placed on the scale insects after they were bathed in oil, they became desiccated and died much more rapidly; none of them survived for a period longer than 10 days.

Reproductive System

The *genitalia* are the external organs of reproduction, with all their appendages. In the females they may be modified into a highly specialized exserted or retractile egg-laying organ or *ovipositor*. It is not present in all orders of insects, but when present it may be used to deposit eggs into the ground, into plant tissues, or into the bodies of animals. The most remarkable modifications of the ovipositor may be found in the Hymenoptera. In the hymenopterous parasites, it is greatly elongated and slender.

The eggs are then compressed and stretched to an extreme degree in their passage through the very narrow channel of the bristlelike ovipositor. In the stinging Hymenoptera (ants, wasps, and bees), the ovipositor is modified into a stinger. The venom is ejected at the tip of the stinger, and the eggs are ejected from the opening of the genital chamber: which lies at the base of the ovipositor.

The female internal reproductive organs consist primarily of a pair of *ovaries*. Each ovary normally consists of several *ovarioles* within which the eggs are developed. From the ovaries issue a pair of *oviducts*, which are usually united and lead to the *vagina*. There is almost always a saclike pouch called the *seminal receptacle* or *spermatheca*, which receives and stores the sperm. Since some insects mate only once, the spermatheca performs an important function. In some insects (e.g., Lepidoptera), there is also a *bursa copulatrix* which receives the sperm first. The *colleterial glands* secrete a viscid material used to "cement" eggs together, waterproof them, attach them to objects, or to form envelopes that may encapsulate a number of eggs.

The number of eggs laid by an insect may vary from a few to many thousand. The queen honey bee may lay 2,000 to 3,000 eggs per day for weeks at a time, and the termite queen can lay millions of eggs. Not all insects lay eggs. Insects which lay eggs are said to be *oviparous*. Their eggs may be fertilized, as they pass through the vagina, by sperm stored in the spermatheca. If the eggs produced by the female bear viable offspring although the eggs are not fertilized, reproduction is said to be *parthenogenetic*. The developing embryo within the egg is nourished by means of the yolk, as in the case of birds.

Unlike the birds, however, insects do not incubate the eggs and, in fact, rarely care for the eggs in any way after they are laid. They usually lay eggs in, on, or near food that will be consumed by the hatched young.

Many species of insects give birth to active, living young. Such insects are said to be *viviparous*. Among such species, the production of young may also be by parthenogenesis as, for example, among the aphids. The term *ovoviviparous* refers to the type of reproduction in which eggs with well-developed shells hatch within the body of the female and active young are then produced. The most recent tendency is to consider that the term "ovoviviparous" is superfluous, because all insects lay eggs, whether these hatch within the body or, as is more often the case, they are laid by the female and hatch outside.

The males of ants, bees, and social wasps are produced from unfertilized eggs and the females and workers from fertilized eggs. The males of some parasitic Hymenoptera and aleyrodids are also produced from unfertilized eggs.

In a diagram of the male and female reproductive organs, the structural analogy can be seen. Comparable to the ovaries are the male testes in which sperm is produced, leading to the *vasa deferentia* which unite to form the *ejaculatory duct*. *Accessory glands* secrete fluid with which the sperms are mixed. The external genitalia consist functionally of the *genital claspers* and the *aedeagus*, which includes the *penis*.

The external genitalia of the male, are used extensively in the description of species because they often show more constant and more distinct differences than any other structural characters to be found on the insect.

Musculature

Insects have a large number of muscles. The external skeleton allows for a large area of attachment for muscles compared to the bony internal skeleton of vertebrates. In man, there are said to be 696 muscles, but Lyonet found over 4,000 in the caterpillar of the goat moth, *Cossus ligniperda*.

The muscles of insects are soft, but quite strong, and they are cross-striated, alternate light and dark bands crossing the fibres. There are *skeletal* and *visceral* muscles. In the latter, the transverse striations are not so distinct as in the skeletal muscles.

Insect muscles are very strong in relation to the insect's body weight. Insects are able to lift from 15 to 25 times their body weight. Their muscles contract and relax with phenomenal rapidity, as illustrated by the great speed of wing vibrations. Insect wing muscle is the most active tissue known, and consumes the most fuel and oxygen per unit of protein. The insect tracheal system facilitates the diffusion of oxygen and carbon dioxide through muscle tissue several thousand times faster than diffusion in the liquid phase.

Metamorphosis

Metamorphosis means "change in form," and with respect to the development of animals, it refers to change in form during postembryonic growth.

In some of the lower forms of insects, there is no change in form; in others, there are varying degrees of change, up to such striking transformations as may be observed between a grub and an adult beetle, a caterpillar and a butterfly, or a maggot and an adult fly.

Development without Metamorphosis

The insects of the 3 orders lowest in the scale of evolution, the Protura, Thysanura, and Collembola, do not go through any change in form after hatching, any more than do the vertebrates. Their development is without metamorphosis (ametabolous development). A young silverfish or springtail, for example, appears so much like the adult that one would immediately infer that the newly hatched young and the adult are the same species. The young insect increases in size, but its shape changes very little. The 3 orders in which no metamorphosis occurs comprise the subclass Apterygota, the insects of which have no wings and are descended from wingless ancestry.

Gradual Metamorphosis

Since no insect has visible wings immediately after birth, any insect that develops wings must undergo some degree of metamorphosis during its life history. Therefore, metamorphosis may be found among all the insects of the subclass Pterygota. These are insects that have wings in the adult stage or, if the wings are absent, they were lost during the course of evolution from winged ancestry. Cockroaches, grasshoppers, earwigs, termites, lice, thrips, bugs, aphids, and scales are representative of the insects that undergo *gradual* metamorphosis (paurometabolous development).

Since the immature stages, commonly called *nymphs*, require about the same kind of habitat and food as the adults, internal anatomical changes during the life history of these insects are not on the same scale as for insects for which food and habitat change drastically. The most conspicuous change other than an increase in size during the development of insects with gradual metamorphosis is the difference in the appearance of the externally developed wing buds.

European entomologists have not been so adamant as those in the United States in using only the term "nymph" for the immature stages of the paurometabolous insects. Some refer to them as "larvae," or may use the term "nymph" for some insects (e.g., cockroaches and earwigs) and the term "larva" for others (e.g., thrips and scale insects). Some use the term "larva" for early instars and "nymph" for later instars of the immature forms of the same paurometabolous insects, such as the termites.

European entomologists did the pioneer work on termite caste development, and used the term "larva" to refer to termite nymphs before they acquire wing buds and the term "nymph" for individuals succeeding the apterous nymphs and possessing external wing buds. For the sake of standardization, American entomologists are using the same terminology for termites.

INCOMPLETE METAMORPHOSIS

Three orders of aquatic insects, the Ephemerida (mayflies), Plecoptera (stoneflies), and Odonata (dragonflies), are placed by some authors in a group called the Hemimetabola, a group having *incomplete* metamorphosis (hemimetabolous development). These insects undergo radical changes in habitat during their life histories.

The nymphs (naiads) and adults of insects with incomplete metamorphosis live in entirely different habitats. The aquatic nymphs have tracheal anal or rectal gills; their legs are modified for clinging, climbing, or burrowing; their bodies are modified for swimming; and their mouthparts are modified for taking food in the water. Special adaptations are required to enable the adults to escape from the last nymphal skin. The adults, being aerial, require different methods for capturing food (dragonflies and damselflies) or they may take little or no food, but inflate the alimentary canal with air to aid in flight.

Obviously, the morphological changes dependent upon a change from aquatic to terrestrial existence must of necessity be greater than those which accompany the changes occurring among insects with gradual metamorphosis, which undergo no change in habitat or food habits.

COMPLETE METAMORPHOSIS

The higher orders of insects, including the 4 that have the greatest number of described species, the Coleoptera (beetles), Lepidoptera (butterflies, moths), Hymenoptera (ants, wasps, bees), and Diptera (flies, mosquitoes), undergo what is known as *complete, complex,* or *indirect* metamorphosis, and are placed in a group called the Holometabola (having holometabolous development).

The young of insects with complete metamorphosis are known as *larvae,* and are so different from the adult that no one unfamiliar with their life history would suspect them of being merely a different stage in the development of the same individual. For example, what is there about the habitat, food habits, and appearance of a codling moth larva, the "worm" of the wormy apple, that would lead an uninformed person to suspect that it is an immature stage of the codling moth? For such radical transformations, the insects with complete metamorphosis have a life stage called the *pupa,* that is not possessed by any of the groups already discussed.

The pupal stage is a quiescent, transient one, in which there is usually little locomotory movement and no feeding. The larva generally pupates in a place where the pupa will be hidden or inconspicuous—the pupa, being motionless, is also defenseless. A protection for the pupa is often spun, or prepared in some other way, by the larva. On the other hand, some dipterous pupae in the Culicidae and Chironomidae are very active, as evidenced by the familiar mosquito "wigglers."

All activity during the pupal stage is concentrated on the complex chemical changes associated with simultaneously breaking down (histolysis) and building up (histogenesis) tissues before the insect can emerge from the pupal skin, transformed for its radically new form and life habits. For example, in the pupa of a moth, sexual organs develop, chewing mouthparts are transformed into coiled, siphoning mouthparts, wings develop, compound eyes and antennae are formed, and in general the insect whose simple, cylindrical larval form was specialized for feeding and storing energy is transformed into a highly active and sensitive organism, specialized for flying about, securing food, finding a mate, and locating a place for oviposition.

One of the advantages of complete metamorphosis is that the larvae are specialized in habits and form for feeding, growing, and storing energy, and the adults are specialized for mating, reproducing, and effecting a dispersal of the species.

The means of locomotion, sense organs, and reproductive organs that the larva did not require are now essential to the adult. In addition, the pupae may provide a means of passing through seasons of adverse weather conditions or periods of food shortage.

Insects with complete metamorphosis develop their wing buds and the buds for all other adult structures internally in the larval instars. The buds often become external in the pupal stage, but are hidden under the pupal sheath in the case of obtect (smooth) pupae, when they are not seen until the adult stage.

Neither the pupa nor the adult ever grows. The layman often supposes that a small house fly is a young fly that has not yet completed its growth. However, differences in the sizes of adults merely indicate variations in the amount of growth attained by the larvae. This, in turn, was largely dependent upon the food supply.

The change of form from larva to adult never fails to excite the wonder and imagination of all people; the development of mammals, including man, appears so simple and uneventful by comparison. In fairness to the mammals, however, it should be pointed out that their greatest changes in form occur during embryonic development--changes which, in fact, recapitulate the greater part of their long and eventful phylogenetic development. The morphologic changes of the higher insects are more spectacular because they occur during the postembryonic development of the animal.

THE TERMS "LNSTAR" AND "STAGE"

Now that some understanding has been gained of the meaning of "metamorphosis," the distinction between the terms *instar* and *stage* can be more readily understood. The cuticle of an insect allows for only limited growth; for further growth, the insect must split and shed its cuticle, and this process is called *molting* or *ecdysis*. Before the old cuticle is shed, a new cuticle is already forming, and will allow for another limited increase in body size. This process usually continues until the insect becomes an adult, although in a few insects such as the proturans (Protura), springtails (Collembola), and silverfish (Thysanura), molting continues even after the adult stage is reached.

An *instar* is the form in which an insect appears after each molt. The period between any 2 molts is called a *stadium* (plural, *stadia*). The life *stages* of an insect are the periods of its life which are radically different from each other in appearance and usually also in behaviour and activity, particularly so in the case of holometabolous insects. For example, a cockroach has 3 stages: egg, nymph, and adult; a moth has 4 stages: egg, larva, pupa, and adult.

Differences Between Nymphs and Larvae

Since the naiads of aquatic insects with incomplete metamorphosis are of comparatively little interest to the majority of entomologists, the accompanying tabulation gives only the factors which differentiate nymphs and larvae.

Some Adult Morphological Specializations

In many of the holometabolous insects, the adults do little feeding; the energy supply for their relatively brief adult life is carried over from the larval stage in the fat bodies. Some females (e.g., mosquitoes and fleas) must feed in order to develop eggs. However, adult specializations are chiefly for reproduction and distribution of the species. In the higher insects, one form of specialization is sexual dimorphism. In some insects, the 2 sexes are so different that they could easily be mistaken for separate species. In others, the male may be partially or completely eliminated, and reproduction is by *parthenogenesis* (e.g., in aphids). In some insects, the normal adult morphological features are absent; characters of the immature stage are carried over to the sexually mature and reproducing individuals. This phenomenon, called *neoteny* is frequently associated with parthenogenetic reproduction and viviparity (rearing active young instead of eggs), and is restricted to the female.

In social insects, different *castes* perform particular functions in the colony. Information on castes and their functions can be found elsewhere n this book in discussions of the biologies of termites, ants, wasps, and bees.

NYMPH

1. Appearance similar to that of adult.
2. Habitat and food same as for adult.
3. Successive nymphal instars become progressively more like adult in appearance.
4. Has no inactive stage between nymph and adult.
5. Wing buds develop externally.
6. Has few, if any, structures different from hose of adult.
7. Mouthparts same type as those of adult.
8. Has compound eyes if compound eyes are possessed by adult.

Larva

1. Appearance radically different from that of adult; typically cylindrical.
2. Habitat and food often greatly different from those of adult.
3. Last larval instar appears no more like the adult than the first instar.
4. Has an inactive pupal stage for transformation from larva to adult.
5. Wing buds develop internally.
6. Has temporary structures, such as prolegs, not retained by adult.
7. Mouthparts often different from those of adult (e.g., caterpillars, chewing; moths, sucking).
8. Never has compound eyes.

Diapause

Common to all forms of life, and apparently as necessary as any other life function, is the phenomenon of rest or *diapause*, involving various forms of cessation of activity. It may take the form of a temporary cessation of development, or temporary cessation of activity, or both.

In insects, diapause most often occurs in the gg or pupal stages, although it may occur in other stages of development. It may be caused by hanges in temperature, moisture, food, water, or oxygen. Diapause may be triggered by a engthening (vernal) photoperiod or by a shortening (autumnal) photoperiod, or may be imposed by such internal factors as heredity, enzymes, or hormones. Some believe that diapause ay be brought about by autointoxication or byhe accumulation of some chemical, somewhat like muscular fatigue in higher animals. Diapause that enables an insect to resist cold is called *hibernation*; if it enables an insect to resist heat or drought, it is called aestivation.

CLASSIFICATION: THE ORDERS OF INSECTS

BINOMIAL NOMENCLATURE

It has been shown that the members of the class exapoda have in common a number of characteristics which definitely separate them from the classes of the phylum Arthropoda. Likewise, the class Hexapoda can be divided into a umber of *orders*, each of which includes insects with certain common characteristics that indicate more or less similar morphological and/or phylogenetic relationships. For example, there are striking points of difference noted in grasshoppers (order Orthoptera), which have 4 wings (the front pair generally thickened), chewing mouthparts, and gradual metamorphosis, as compared with the flies (order Diptera), which have a single pair of wings, sponging or piercing mouthparts, and complete metamorphosis. The various orders can usually be readily distinguished, even by relatively inexperienced students of entomology.

The orders are subdivided into families, the families into genera (sing. genus), the genera into species, and the species sometimes are subdivided into subspecies or varieties. The accompanying classification of man in comparison to that of one of his external parasites, the head louse, *Pediculus humanus capitis* De Geer, will illustrate the significance and the relative magnitude of the various categories universally accepted in both plant and animal classification. However, only a relatively few species of insects have been divided into taxonomically distinct subspecies.

	Man	**Head Louse**
Kingdom	Animal	Animal
Phylum	Chordata	Arthropoda
Class	Mammalia	Hexapoda
Order	Primates	Phthiraptera
Family	Hominidae	Pediculidae
Genus	Homo	Pediculus
Species	sapiens	humanus
Subspecies	Caucasian	capitis

In addition to the categories shown in the box, additional subdivisions have been established to facilitate the classification of certain groups whose large size or structural peculiarities make further division desirable. For example, orders may be divided into suborders, a number of families may be grouped into a superfamily, or a family may be divided into a number of subfamilies.

A subfamily may in turn be divided into tribes, each consisting of a number of genera. The suffix for the name of the superfamily is "-oidea." This should be distinguished from "-odea," which is used as a suffix for the names of some of the orders and suborders. The suffix for the name of the family is "-idae," for the sub- family, "-inae," and for the tribe, "-ini." Thus, one of the familiar tarantula hawks of the southwestern United States, *Pepsis formosa,* is in the superfamily Vespoidea, family Pompilidae, subfamily Pepsinae, and tribe Pepsini

Some people are inclined to think of scientific names as pedantic gestures, or of having perhaps some vague academic significance of small concern to anyone besides the taxonomist.

The common names of most animals, especially insects, are highly variable, even within a relatively limited area, and may mean nothing to people in other areas within the same country, to say nothing of those in other parts of the world. Likewise, common names may give little or no indication of the taxonomic relationship of the insects another. Therefore, only the scientific have a universal significance—referring same plant or animal the world over.

In the Entomological Society of America, there is a Committee on Common Names of Insects. This committee has developed a list of approved common names of insects and a few other arthropods, and also their scientific names, according to the latest concepts of taxonomists in each insect group. This list is periodically revised, expanded, and sent to members of the society. By following the recommendations of the committee, American and Canadian entomologists are standardizing nomenclature in their published papers.

Binomial Nomenclature

Before the time of Linnaeus, the name of the insect was a short Latin description. It can be imagined how cluttered with names entomological literature would be if such a system of nomenclature had been retained. Linnaeus suggested that the name of an animal (or plant) be composed of the compounded names proposed for the genus and for the species of the animal in question. These were always Latin names, for Latin was the universal language among scholars of that time. Latin has an additional value for use in "scientific names," in that its form and usage are now stereotyped, and can be used with unchanged meaning by people of all languages and in all parts of the world.

We hear much nowadays about the desirability of a universal language. A universal language has been used for centuries by biologists in the naming of plants and animals, and if this had not been the case, the chaotic condition of biological literature would be difficult to imagine. Certainly, natural science would have been greatly retarded. Chemists are also in world-wide agreement as to chemical nomenclature, having adopted a universal system at a conference held at Geneva, Switzerland, in 1892.

Linnaeus grouped the insects into 7 orders: Coleoptera, Hemiptera, Lepidoptera, Neuroptera, Hymenoptera, Diptera, and Aptera. He described and named a surprisingly large number of species, as a perusal of general entomological literature will show. In 834 pages, Linnaeus described 4,379 animals, including 1,937 insects. By now, about a million species of insects have been named and described, and about 6 or 7 thousand new species are being added in the world's taxonomic literature each year. There are probably at least as many species of insects still unnamed and undescribed.

Linnaeus perfected the binomial system of classification, and used it in his famous general treatise on animals, called *Systema Naturae*. The tenth edition of this publication, which appeared in 1758, is universally accepted as the starting point in binomial nomenclature. Any specific name proposed before the date of publication of the tenth edition of *Systema Naturae* is not considered valid. In all the naming of species subsequent to 1758, however, the rules of priority are strictly followed. If a species is described and named, and subsequent research reveals that someone else had already described and named the same insect, the later name becomes invalid. Taxonomists are no more infallible than any other specialists, and the constant revision of initially presumed taxonomic relationships of species, genera, and families of insects to one another is a task that continues apace with the description of new species.

For complete information on a scientific name, the name of the "author," the researcher who first described and named the species, follows the name of the species, but his name is not underscored or written in italics. In most entomological literature, the author's name is abbreviated. However, the modern tendency is to write out the author's name completely, since there are now too many names to abbreviate without duplication or confusion. Likewise, beginning students and those not familiar with a given insect group are not likely to know the meaning of the majority of the abbreviations. However, an abbreviation for Linnaeus or Fabricius is acceptable.

The house fly may be taken as an example to illustrate the correct form of binomial nomenclature. It is one of the many species which was named by Linnaeus. The scientific name of the house fly is *Musca domestica* Linnaeus (or *Musca domestica* Linn., or L.). One knows from the scientific name that Linnaeus described the house fly as the species *domestica* and that he placed in the genus *Musca*. Let us take another example, the codling moth of apples, pears, and walnuts, another well-known cosmopolitan insect pest.

The scientific name of this insect is *Laspeyresia pomonella* (Linnaeus). The parentheses around "Linnaeus" show that he named the codling moth *pomonella*, but that he placed it in some other genus (*Tinea*). Later, a specialist in this particular group of the Lepidoptera found that the codling moth showed greater taxonomic relationships to insects which belong in the genus *Laspeyresia*, a genus that had not yet been recognized in Linnaeus' time. Since then, the species *pomonella* has actually been placed in 3 successive genera, *Cydia*, *Carpocapsa*, and *Laspeyresia*, the latter being currently recognized by entomologists and approved by the Entomological Society of America.

Relation of Scientific Names to Entomological Investigation

The first step in investigating an insect problem is to establish the identity (scientific name) of the insect in question. If the name of the insect is not known to the investigator, or if he is not absolutely certain of its identity, he may send specimens to the United States National Museum for identification, or if he knows to what group (family or genus) the insect belongs, he may send specimens directly to a taxonomist who specializes in that particular group of insects.

When he is certain of the scientific name of the insect, the investigator proceeds to examine the entomological literature for information on it.

The literature may reveal to an investigator much that is already known about an insect, such as its economic importance, geographic distribution, hosts, life history, seasonal history, habits, tropisms and ecological relationships, climatic or other physical factors favourable or unfavourable to its abundance, natural enemies, and previous investigations on control measures.

The obtaining of all this wealth of information is predicated on the correct identification of the insect species and the ability to use the world's entomological literature on the basis of a standardized system of nomenclature. No matter in what country the literature originated, the insect should always have the same scientific name, for insects are named according to universally accepted standards of nomenclature. Without a universally accepted system of classification and naming of insects, no other phase of entomological investigation could be systematically pursued.

Some Characteristics of the Phylum Arthropoda

As shown in the various categories of classification, the phylum is the most inclusive of the subdivisions of the animal kingdom. In the Arthropoda, the various classes (subdivisions of a phylum) share these characteristics in common: bilateral symmetry; segmented bodies; paired, jointed appendages; protective cuticle; a dorsal blood vessel; and a ventral nervous system. Some, but not all, of these characteristics are possessed by some groups of animals farther down in the scale of evolution. For example, the members of the phylum Annelida, to which the earthworms belong, possess bilateral symmetry; segmented bodies; protective cuticle; a dorsal blood vessel; and a double ventral nerve cord, segmented into a series of ganglia. However, they do not have jointed appendages.

No other phylum of animals possesses *all* of the characteristics of Arthropoda. Within the phylum Arthropoda, however, there are some points of difference which relegate different groups of animals to different classes. The class Hexapoda, for example, is the only group of arthropods that has 3 body regions, 3 pairs of legs, and wings (as adults).

Identification of Insects

When an insect taxonomist or systematist is given a large group of insects to separate into their respective orders, families, genera, and species, he first attempts to sort out the species he recognizes from earlier experience. For aid in determining the relationships of the insects of whose identification he is not certain, the taxonomist will often use "keys," by means of which the insects can be traced through a successive series of morphological characters. The keys are so arranged as to facilitate the determination of scientific names and systematic relationships, and have been prepared for all the categories of insect classification. They are based on the work of a large number of specialists in the various groups of insects, and are, of course, subject to revision as new information is gained regarding the taxonomic relationships of the various groups. The use of a key depends on some knowledge of the structural characteristics and terminology of the group of animals (or plants) for which it has been worked out.

The accompanying simple and admittedly incomplete key was designed to enable students to "key out" the principal orders of adult insects for a course in economic entomology. The orders Protura and Collembola are included among the insects, although modern taxonomists consider them to belong to different subclasses. In this simplified key the diplurans (Diplura) are considered to be in the order Thysanura and the cockroaches are considered to be in the order Orthoptera, although the cockroaches are placed in Dictyoptera. The domestic species of the cockroaches, flies, and ants. These are insects whose habits are such that they commonly come in contact with food or crawl about on surfaces on which it is placed, such as shelves, sinks, and breadboards. Cockroaches and flies are known to be mechanical carriers of many kinds of disease germs.

Index

D

E

F

G

H

I

J

L

M

N

O

P

Q

R

S

T

U

V

W

X

Z